Assessment of Vulnerability to Natural Hazards

Assessment of Vulnerability to Natural Hazards

A European Perspective

Edited by

Jörn Birkmann
United Nations University,
Institute for Environment and Human
Security (UNU-EHS), Bonn, Germany

Stefan Kienberger
Department of Geoinformatics –
Z_GIS, University of Salzburg,
Salzburg, Austria

David E. Alexander
Institute for Risk and Disaster Reduction,
University College London, London,
United Kingdom

ELSEVIER

AMSTERDAM • BOSTON • HEIDELBERG • LONDON
NEW YORK • OXFORD • PARIS • SAN DIEGO
SAN FRANCISCO • SINGAPORE • SYDNEY • TOKYO

Elsevier
525 B Street, Suite 1900, San Diego, CA 92101-4495, USA
225 Wyman Street, Waltham, MA 02451, USA

Library of Congress Cataloging-in-Publication Data
Assessment of vulnerability to natural hazards: a European perspective/[edited by] Jörn Birkmann, Stefan Kienberger, David E. Alexander.
pages cm
ISBN 978-0-12-410528-7 (hardback)
1. Natural disasters–Europe. 2. Hazard mitigation–Europe 3. Environmental risk assessment–Europe. 4. Emergency management–Planning–Europe. 5. Human security. I. Birkmann, Jörn. II. Kienberger, Stefan. III. Alexander, David (David E.)
GB5008.E87A77 2014
363.34'2064–dc23
2014006590

British Library Cataloguing in Publication Data
A catalogue record for this book is available from the British Library

ISBN: 978-0-12-410528-7

For information on all Elsevier publications visit our web site at store.elsevier.com

Printed and bound in China

14 15 16 17 18 10 9 8 7 6 5 4 3 2 1

Contents

Contributors

David E. Alexander Institute for Risk and Disaster Reduction, University College London, London, United Kingdom

Marjory Angignard Institute of Spatial Planning, Technical University of Dortmund, Dortmund, Germany

Alex H. Barbat Universitat Politècnica de Catalunya, Centre Internacional de Mètodes Numèrics en Enginyeria (CIMNE), Barcelona, Spain

Jörn Birkmann United Nations University, Institute for Environment and Human Security (UNU-EHS), Bonn, Germany

Omar D. Cardona Universidad Nacional de Colombia, Sede Manizales, Colombia

Martha Liliana Carreño Centre Internacional de Mètodes Numèrics en Enginyeria (CIMNE), Universitat Politècnica de Catalunya, Barcelona, Spain

Dr Salete Carvalho Department of Geography, Faculty of Arts, University of Porto, Porto, Portugal

Diana Contreras Department of Geoinformatics – Z_GIS, University of Salzburg, Salzburg, Austria

Yaella Depietri United Nations University, Institute for Environment and Human Security (UNU-EHS), Bonn, Germany; Institut de Ciècia i Tecnologia Ambientals (ICTA),Universitat Autònoma de Barcelona (UAB), Barcelona, Spain

Dr Nicolas Desramaut BRGM, Risks and Prevention Division, Ground Instabilities and Erosion Risk Unit, Orléans, France

Unni Eidswig NGI, Oslo, Norway

Dr Manuel Garcin BRGM, Risks and Prevention Division, Coastal Risks and Climate Change Unit, Orléans, France

Thomas Glade Department of Geography and Regional Research, University of Vienna, Vienna, Austria

Stefan Greiving Institute of Spatial Planning, Technical University of Dortmund, Dortmund, Germany

Christian Iasio Institute for Applied Remote Sensing, EURAC European Academy for Research, Bolzano, Italy

Margareth Keiler Institute of Geography, University of Bern, Bern, Switzerland

Stefan Kienberger Department of Geoinformatics – Z_GIS, University of Salzburg, Salzburg, Austria

Maria Papathoma-Köhle Department of Geography and Regional Research, University of Vienna, Vienna, Austria

Mabel C. Marulanda Centre Internacional de Mètodes Numèrics en Enginyeria (CIMNE) Barcelona, Spain

Roberto Miniati Department of Information Engineering, University of Florence, Florence, Italy

Lydia Pedoth Institute for Applied Remote Sensing, European Academy of Bolzano (EURAC), Bozen/Bolzano, Italy

Mark Pelling Department of Geography, King's College London, The Strand, London, United Kingdom

Fabrice Renaud United Nations University, Institute for Environment and Human Security (UNU-EHS), Bonn, Germany

Dr Jeremy Rohmer BRGM, Risks and Prevention Division, Risks of underground Storages and Exploitations, Orléans, France

Stefan Schneiderbauer Institute for Applied Remote Sensing, European Academy of Bolzano (EURAC), Bozen/Bolzano, Italy

Reinhold Totschnig eb&p Umweltbüro GmbH, Klagenfurt, Austria

Dr Fantina Tedim Department of Geography, Faculty of Arts, University of Porto, Porto, Portugal

Thorsten Ulbrich Institute of Meteorology, Department of Earth Sciences, Freie Universität Berlin

Dr Charlotte Vinchon BRGM, Risks and Prevention Division, Coastal Risks and Climate Change Unit, Orléans, France

Torsten Welle United Nations University, Institute for Environment and Human Security (UNU-EHS), Bonn, Germany

Zehra Zaidi Department of Geography, King's College London, The Strand, London, United Kingdom

Peter Zeil Department of Geoinformatics – Z_GIS, University of Salzburg, Salzburg, Austria

Introduction

Vulnerability: a key determinant of risk and its importance for risk management and sustainability

Jörn Birkmann, Stefan Kienberger, David E. Alexander

Bush fires in February 2014 in Norway, extreme heat stress in 2013 in London, and major floods in June 2013 in Central Europe illustrate that communities and states in Europe have to prepare seriously for extreme events and natural hazards. While some phenomena are unexpected, such as the bush fires in Norway during winter, other hazards are more or less well known, such as floods along large river systems. However, similar hazards and extreme events often have quite different impacts and consequences. For example the major earthquake and tsunami in Japan in March 2011, with a magnitude of 9.0 M_w (moment magnitude scale) caused about 18,500 fatalities, while the earthquake disaster in Haiti in 2010, which had a magnitude of 7.0 M_w—hence about 100 times less powerful than the Tohoku earthquake—resulted in more than 220,000 deaths.

Also in Europe, various population groups were exposed to heat stress in 2003, with the loss of between 22,000 and 70,000 deaths (Robine et al., 2008). Mortality and morbidity were particularly high among the elderly. These considerations lead to several important questions:

- What factors determine risk?
- Why did the earthquake in Haiti, with a significantly lower magnitude cause many more fatalities compared to the event in Japan?
- Why did so many elderly people died during the European heat wave in 2003?
- Which factors have significantly increased the risk of harm and loss of life or property due to extreme events and natural hazards?
- Can we measure differences in vulnerability and capacity to respond before such extreme events strike societies?
- Can these assessments help to identify relevant intervention measures designed to mitigate risk?

This book aims to provide some answers to the questions listed above. It focuses on the development and application of vulnerability assessments to natural hazards in Europe. Vulnerability assessment today is a key endeavor within different scientific communities, such as disaster risk reduction, emergency management, and climate change adaptation. In this regard, there is an emerging consensus that extreme events and natural hazards do not necessarily cause extreme impacts and major harm, but rather that the vulnerability of a society, community, or system (infrastructure, social-ecological system, etc.) exposed to the hazard determines whether it translates into disaster.

The concept of vulnerability has been used in the English language for 400 years, although its origins in Latin (*vulnerare*, to damage or wound) go back to time immemorial. Vulnerability implies an inherent or innate susceptibility to harm and is the antithesis of robustness, resilience, and ability to resist. In terms of human process, it is often seen as the opposite of ability to cope or to bring to bear sufficient capacity to overcome and neutralize harm when it threatens a person or entity. Many more precise definitions of vulnerability have been developed in the various contexts in which it has been used. As a result, in the study of extreme events and disasters it has acquired a variety of meanings and overtones whose proliferation and contrasts have led to a degree of confusion over definitions. For example, does it include fragility, susceptibility, and exposure, or are these separate concepts? Nevertheless, in disaster studies, vulnerability has gradually established itself as a central concept, and many would regard it as the key to understanding the impact of extreme events. To produce disaster, hazards act upon the vulnerable fabric of society, and the vulnerable environment in which society has its being. An extreme but popular view is to treat hazard as merely the trigger of disaster and to argue that society is so complex that the ramifications of its vulnerability, and associated feedback loops, largely determine the form and magnitude of disaster. Another way of looking at this is that disaster cannot be understood, and hence neither can disaster risk, without an intimate knowledge of vulnerability in the various facets and categories in which it is manifest in society and environment.

Under conditions of global environmental change, it is expected that Europe will face increases in the intensity and frequency of extreme natural events. At the same time important changes in societal conditions and preparedness can be observed in countries and regions in Europe. For example, the demographic change in Germany will most likely lead to a larger number of elderly people, which will make many regions and their populations in Germany more vulnerable to heat stress, particularly as elderly people have only limited physical means to cope with heat stress. In addition, socioeconomic changes and modifications in employment situations or social security networks, such as in Greece, might also influence the vulnerability of people who are exposed to natural hazards such as forest fires. Finally, increasing urbanization of hazardous areas, high levels of dependency on critical infrastructure (electricity, gas, water, banking, etc.), and the increasing vulnerability of particular social groups point to the urgent need to improve, not only our knowledge about the physical phenomena and hazard characteristics, but also vulnerability and its underlying factors in Europe and among its citizens.

GOALS OF THE PROJECT

This book is based on the outcome of a research project funded by the European Commission's Framework Project FP7. The project was named MOVE (Methods for the Improvement of Vulnerability Assessment in Europe) and it was designed to:

- Enhance the base of knowledge on frameworks and methods for the assessment of vulnerability to natural hazards in Europe.
- Use indices and indicators to help improve societal and environmental resilience by placing emphasis on clear, capable measurement and by accounting for uncertainties.
- Identify gaps in existing methodologies for accomplishing these tasks.
- Produce a conceptual framework that is independent of scale and hazard type.
- Analyze physical, technical, environmental, economic, social, cultural, and institutional vulnerability measured for specific hazards and at different geographical scales.
- Study the vulnerability of people and specific infrastructures (e.g., hospitals) to floods, temperature extremes, droughts, landslides, earthquakes, wildfires, and storms impacts.

While the development of a common framework was an important stimulus to discussion and the harmonization of different definitions of vulnerability, the empirical case studies showed that, next to a common metaframework, the characteristics of each case study are specific to its context. The information that they contributed to the study might require the modification of the overall framework. In this respect, the empirical research on the ground allowed to operationalize the different components and factors of vulnerability outlined in the framework for the specific case studies.

The project involved substantial participation by stakeholders, who were comprehensively consulted. This provided a basis to explore how these new methods and the information gathered within the vulnerability assessments can be linked to existing evaluation and planning tools. For example, the framework and case studies on flood vulnerability have a strong relevance for policy makers who aim to create a comprehensive and integrated approach to decision making for risk identification and vulnerability reduction. The European Union (EU) Flood Directive on Assessment and Management of Flood Risks (2007/60/EC), and the subsequent EU Flood Directive Implementation Strategy, lack the ability to address vulnerability in a comprehensive and integrative manner. In this regard, the MOVE framework and the practical results of the vulnerability assessments of floods demonstrate that flood risk mapping and management will have to consider different dimensions and factors of vulnerability if the aim is to manage and reduce flood risk effectively in Europe.

Furthermore, case studies in Germany, Italy, Spain, Austria, the United Kingdom, and Portugal show that vulnerability assessment should not be limited to economic loss estimation, as it also requires the consideration of tangible and intangible factors that determine institutional and cultural vulnerability.

Hence, the core part of the book is designed to provide essential insights into case studies on how the proposed MOVE framework can be operationalized. The context and the settings of the case studies are extremely diverse, and so are the methods applied, results achieved, and conclusions gained. Urban case studies (such as those conducted in Barcelona and London) demonstrate the complexity of urban environments, but they also highlight the need to achieve a detailed understanding of underlying causal and spatial patterns. Focusing on a range of hazard types (as did the Cologne and South Tyrol case studies) highlights the difficulties of creating a common framework in an environment of diverse multiple hazards, and it emphasizes the challenge of providing the tools to identify possible "generic" vulnerability factors. Innovative methods range from the modeling of vulnerability regions in a spatially explicit manner that is independent of administrative units (as shown in the Salzach River case study) to the assessment of institutional vulnerabilities (as evinced in the Cologne case study). Moreover, specific assessments of the impact of hazards on health-care systems (as in the Florence case study), coastal erosion, and forest-fire-related assessments (as in case studies from Portugal) point the way toward more holistic assessments (such as the Cologne and Salzach case studies) and demonstrate the variety of ways in which vulnerability can be measured under a common framework.

Overall, this volume can provide important guidance and methodology on how to translate the complex concept of vulnerability into the practical assessments that are essential to the identification of risks and to holistic risk management approaches. Understanding vulnerability as a crosscutting concept that links disaster risk management, climate change adaptation, and development processes. It can contribute much to the development of integrated policies. In this regard, it is important to note that the nexus between development, risk reduction, and climate change is increasingly receiving attention in European and international policy making. In the ongoing climate change negotiations—particularly in terms of the subprogramme on loss and damage—specific tools and monitoring systems will be developed to assess progress and reduce the risk of loss and damage in the context of both the extreme events and creeping changes that are linked with climate change. The definition of the new Sustainable Development Goals and the further development of the Hyogo Framework for Action in the post-2015 process require assessment methods to be enhanced and indicators to be developed in order to measure progress toward the chosen goals and targets. In this broader context of scientific discourse and policy, this book will provide important input to an improved understanding of essential factors that might determine whether a natural hazard or event can lead to a disaster or crisis.

ACKNOWLEDGMENTS

We would like to thank the practitioners who supported the research and dissemination work conducted within the MOVE project. In addition, a sincere word of thanks go to the external reviewers and experts of the MOVE project and the different chapters of the volume, particularly to Christian Kuhlicke and John Twigg, Melanie Gall, Hugh Deeming, and Denis Chang Seng for their valuable comments and critique. Furthermore, we are very grateful for the funding received for the project by the European Commission DG Research and the continuous advice received from Mr. Denis Peter. Last, but not least we would also like to thank the Editorial team of Elsevier, especially Louisa Hutchins and Sharmila Vadivelan who provided us with effective and continuous support in terms of the management of the publication process.

REFERENCE

Robine, J.-M., Cheung, S.L.K., Le Roy, S., Van Oyen, H., Griffiths, C., Michel, J.-P., Herrmann, F.R., 2008. Death toll exceeded 70,000 in Europe during the summer of 2003. C. R. Biol. 331 (2), 171–8.

Chapter 1

Theoretical and Conceptual Framework for the Assessment of Vulnerability to Natural Hazards and Climate Change in Europe[1]

The MOVE Framework

Jörn Birkmann*, Omar D. Cardona†, Martha Liliana Carreño§§, Alex H. Barbat¶¶, Mark Pelling‡, Stefan Schneiderbauer§, Stefan Kienberger¶, Margareth Keiler, David E. Alexander††, Peter Zeil¶ and Torsten Welle***

**United Nations University, Institute for Environment and Human Security (UNU-EHS), Bonn, Germany, †Universidad Nacional de Colombia, Sede Manizales, Colombia, ‡Department of Geography, King's College London, The Strand, London, §Institute for Applied Remote Sensing, EURAC, Bozen/Bolzano, Italy, ¶Department of Geoinformatics – Z_GIS, University of Salzburg, Salzburg, Austria, **Institute of Geography, University of Bern, Bern, Switzerland, ††Institute for Risk and Disaster Reduction, University College London, London, United Kingdom, ¶¶Universitat Politècnica de Catalunya, Centre Internacional de Mètodes Numèrics en Enginyeria (CIMNE), Barcelona, Spain, §§Centre Internacional de Mètodes Numèrics en Enginyeria (CIMNE), Universitat Politècnica de Catalunya, Barcelona, Spain*

Chapter Outline

1. This chapter is based on a paper published in Natural Hazards dealing with the MOVE framework; see in detail Birkmann et al., 2013.

Assessment of Vulnerability to Natural Hazards. http://dx.doi.org/10.1016/B978-0-12-410528-7.00001-1

1.1 INTRODUCTION

Newest global assessment reports (GAR, 2011; Welle et al., 2012; IPCC, 2013; IPCC, 2014) underscore that risk reduction and resilience building remains a key challenge for developing and developed countries alike particularly due to the increasing exposure of people and assets in high risk zones and the intensification of extreme events in the context of climate change (see e.g., IPCC, 2013). It is increasingly recognized that natural hazard associated risk and threats to human security cannot be reduced by focusing solely on the hazards. Societies will have to live with changing environmental conditions and therefore need to build resilience by reducing vulnerabilities to natural hazards. Vulnerability assessment of natural hazards and climate change has emerged in the past decades as an important research field (see e.g., Maskrey, 1984; Chambers, 1989; Pelling, 1997; Cardona, 2001; Birkmann, 2006a,b; Adger, 2006; IPCC, 2007; Bohle, 2008; Bohle and Glade, 2008; Oxfam America, 2009; Birkmann, 2013) bringing together scientists from different disciplines (Fuchs, 2009). The following chapter outlines a framework for multidimensional, holistic vulnerability assessment that is understood as part of risk evaluation and risk management in the context of Disaster Risk Management (DRM)[2] and Climate Change Adaptation (CCA). As a heuristic, the framework is a thinking tool to guide systematic assessments of vulnerability and to provide a basis for comparative indicators and criteria development to assess key factors and various dimensions of vulnerability, particularly in regions in Europe, however, it can also be applied in other world regions. The framework has been developed within the context of the research project MOVE (Methods for the Improvement of Vulnerability Assessment in Europe; www.move-fp7.eu) sponsored by the European Commission within the framework of the FP 7 program.

2. The term Disaster Risk Management (DRM) also encompasses the concept of Disaster Risk Reduction (DRR). Thus these terms are used almost synonymously.

1.2 RISK, VULNERABILITY, AND ADAPTATION TO NATURAL HAZARDS

1.2.1 Definitions and Perspectives

The concept of vulnerability is today a core concept that links important research communities, particularly DRM, CCA, and development research. However, definitions of vulnerability are contested and research on vulnerability is underpinned by multiple disciplinary theories based upon natural or social science epistemologies. This results in a range of paradigms for approaching vulnerability and is supported by qualitative and quantitative assessment methodologies (Pelling, 2001; Birkmann, 2006a,b; O'Brien et al., 2007; Birkmann, 2013; Fuchs, 2009).

The natural science research communities often focus on the quantification of different factors of vulnerability (e.g., Kienberger et al., 2009; O'Brien et al., 2007). The aim of these approaches—particularly regarding physical vulnerability—is to define and quantify damage ranges illustrated through vulnerability curves in order to help determine acceptable levels of potential losses (Papathoma-Köhle et al., 2011). Social science approaches often encompass a broad focus and examine, in particular, the likelihood that an individual household or a community will suffer harm or experience losses related to environmental hazards, as well as the context conditions that influence social vulnerability (DFID, 1999; Wisner et al., 2004). In this context, Phillips and Fordham (2009) emphasize that social vulnerability to natural hazards is driven by social inequality and is deeply embedded in social structures that are often resistant to change (Phillips and Fordham, 2009). Besides a clear differentiation of risk and vulnerability, the MOVE framework aims at integrating the concept of adaptation in vulnerability assessments to natural hazards. In this way, the MOVE framework seeks to enhance the DRR perspective by integrating new understanding of coupling, adaptation, and resilience. In this regard, the framework might also serve as a tool to link particularly the DRR and CCA community as well as the resilience research community.

Before discussing the different components of the framework and the key factors of vulnerability in more depth, it is essential to outline the basic understanding of disaster risk and vulnerability as well as adaptation applied within the context of the framework.

1.2.1.1 Social Construction of Risk

The concept of vulnerability underscores the social construction of risk and is supported empirically by a range of studies applying vulnerability to help understand risk to hazards, including those with a focus on climate change (Aysan, 1993; Blaikie et al., 1996; Wisner et al., 2004). Vulnerability refers to the propensity of exposed elements such as physical or capital assets, as well as human beings and their livelihoods, to experience harm and suffer damage and loss when impacted by single or compound hazard events (UNDRO, 1980; Timmerman, 1981; Maskrey, 1984; Cardona, 1986, 1990; Liverman, 1990; Cannon, 1994, 2006; Blaikie et al., 1996, UNISDR, 2004, 2009; Birkmann, 2006b,c; Cutter et al., 2003; Cutter and Finch, 2008; Cutter et al., 2008). While there is broad agreement on the aspects of the social construction of risk (Chambers, 1989; Cannon, 1994; Lavell, 1999; Wisner, 2006; Carreño et al., 2007a,b; Cardona, 2004), it is at the level of measurement where the challenge remains. The difficulty lies in assessing the various dimensions of vulnerability and its multifaceted and dynamic nature (see e.g., Birkmann, 2006b,c; Birkmann, 2013).

1.2.1.2 Risk and Disasters

While risk and vulnerability can be seen as continuums, a disaster is but a moment or materialization of these underlying conditions. Dynamic changes of vulnerability and hazard phenomena also mean that risk is nonstatic; it changes over time and these changes have to be considered when applying specific assessments, as well as when developing corrective (current risk) or prospective (future risk) interventions. Overall, disasters are the product of a complex relationship between the physical environment, both the natural and built environment, and society; its behavior, function, organization, and development, including human perception (Quarantelli, 1998). The term disaster often refers to a social condition whereby the normal functioning of a social system has been severely interrupted by the levels of loss, damage, and impact suffered (Cardona, 1990; Alexander, 1993, 2000; Birkmann, 2006b). However, disaster can also function as a catalyst for change (see e.g., Birkmann et al., 2010). Pelling and Dill (2010) demonstrate how disasters and crises can also catalyze reorganization and learning processes in communities or societies, often accelerating underlying policy and social trajectories. The concept of resilience has developed in different schools of thought, such as ecology (e.g., Holling, 1973), psychology (e.g., Bonanno et al., 2006; Bonanno, 2008), social-ecological systems research (e.g., Berkes et al., 2003; Folke, 2006), and critical infrastructures (e.g., Boin and McConnell, 2007). In general, resilience research is concerned with the ability of a system or a person to deal with disturbances and the effect of stressors. In addition, resilience research, particularly with regard to social-ecological systems or infrastructures, focuses on capacities of systems to reorganize themselves in the face of adverse events through processes described as revolt and remember and respective innovation processes. Within the framework, we refer to resilience mainly in terms of the lack of resilience, hence the limited capacities to cope or to recover in the face of adverse consequences. At the same time we acknowledge that "improving resilience" is part of adaptation within our framework. Hence the resilience concept is attributed to two core components of the MOVE framework: coping and adaptation.

1.2.1.3 Adaptation and Coping

Adaptation in early reports of the IPCC has not received much attention, since the overall perception was that too much emphasis on adaptation would contradict strict goals for the reduction of green house gas emissions. However, today adaptation is a core concept with the climate change context and goes far beyond a rather biological and reactive understanding of adjustment. Hence, adaptation presents itself as a continuous property, with levels of adaptive capacity changing over time as the status of vulnerability components identified above change and the demands of a shifting risk environment alter the appropriateness of particular asset bundles for risk reduction (Pelling, 2001). Such changes can be a result of disaster events but also everyday processes of development. It should be noted that adaptation is distinct from coping. Coping we see as an aspect of resilience that signifies the "here and now" capacity and includes a set of actions available to those at risk. Coping in this way is part of the formula that determines vulnerability at any one moment in time. However, coping mainly deals with the conservation and protection of the current system and institutional settings (see Birkmann, 2010). Adaptation, by contrast, denotes a longer-term and constantly unfolding process of learning, experimentation, and change that feeds into vulnerability. Adaptation can be felt acting to shape all aspects of vulnerability and is observable through the systems and outcomes of learning—planned and spontaneous, pre and postdisaster (Pelling, 2010). This understanding of adaptation is commensurate with the emerging consensus from climate change (see Kelly and Adger, 2000; Yohe and Tol, 2002; Pelling, 2010) where coping is seen to describe actions taken within existing constraints (including vision and knowledge), and adaptation signifies change in the framing institutions.

1.2.2 Frameworks on How to Systematize Vulnerability in Different Communities

In recent years different frameworks have been developed to systematize risk, different facets of vulnerability and resilience (see overview in Birkmann, 2013). Thomalla et al. (2006) and Mitchell and van Aalst (2009) examine commonalities and differences between the CCA and DRM communities and identify key areas of convergence. They conclude that the two communities perceive the nature and timescale of the threat differently: if impacts due to climate change are surrounded by uncertainty, considerable knowledge and certainty exists about the event characteristics and exposures related to extreme environmental conditions based on historical experience. However, it has to be acknowledged that climate change challenges the historical knowledge of natural hazard events, particularly due to the modification of frequency and intensity of such events (Keiler et al., 2010).

CCA increasingly places emphasis on improving the capacity of governments and communities to address existing vulnerabilities to current climate variability and climatic extremes (Thomalla et al., 2006). Echoing the long-standing concerns of the disaster management community for a more socially informed approach to risk management (e.g., Hewitt, 1983; Burton et al., 1993) and from the climate change community, O'Brien et al. (2004) call for an integration of "underlying causes" of vulnerability and adaptive capacity in climate change impact assessments rather than only focusing on adaptive capacity and technical measures. Furthermore, Birkmann and Teichman (2010) argue that the DRM and the CCA communities differ particularly in terms of the spatial, temporal, and functional scales applied within their research. Additionally, Romieu et al. (2010) examined different frameworks and assessment approaches used within CCA and DRM. They concluded that differences are particularly linked to process (stress versus shock), scale (temporal, functional, and spatial), assessment approach (statistical versus prospective), and levels of uncertainty.

1.2.2.1 Different Frameworks to Systematize and Define Vulnerability

The DRM and CCA communities share common roots in social and political science, however, four distinct approaches to understanding vulnerability and risk can be identified. The four approaches are not contradictory but rather approach risk from a specific viewpoint and with particular ends in mind—from the unearthing of systems linkages from the global to the local to the search for quantifiable risk measures. This section provides a brief overview of each approach to help illustrate the key differences and similarities behind these ways of conceptualizing and measuring risk and its components. Each approach has been considered in the production of the integrated framework proposed in this paper. The four approaches can be distinguished between those that are rooted in (1) political economy; (2) social-ecology; (3) vulnerability and disaster risk assessment from a holistic view; and (4) climate change systems science.

The political economy approach can be illustrated by the pressure and release (PAR) model published in Blaikie et al. (1994) and Wisner et al. (2004). This links vulnerability to unsafe conditions and discrete risk in a continuum of vulnerability that connects local risk to wider national and global shifts in the political economy of resources and political power. Associated with this approach and operating across development studies more generally is the Sustainable Livelihood Framework (see DFID, 1999). Applied in risk contexts most commonly to help understand household impacts and coping when faced with food insecurity, the framework successfully unpacks the range of assets that are at risk and can be used to generate security from disaster. Importantly from a political economy perspective, the framework directs attention to the ways in which the organizational, institutional, and political context helps to shape local capacity—but also recognizes that these structures are reproduced through the actions of individuals and households. The social-ecology perspective emphasizes the need to focus on coupled human-environmental systems when dealing with the assessment of risk. The best known visualization of this approach has been developed and published by Turner et al. (2003). Compared to political economy, the perspective of social ecology stresses the transformative qualities of society with regard to nature—and also the effects of changes in the environment on social and economic systems. It argues that the exposure and susceptibility of a system can only be adequately understood if these coupling processes and interactions are addressed.

Comprehensive perspectives from vulnerability and disaster risk assessment have tried to develop an integrated explanation of risk. These approaches particularly differentiate exposure, susceptibility, and societal response capacities or the lack of resilience (see Cardona, 1999a,b, 2001, 2010; IDEA, 2005; Birkmann, 2006a; Carreño, 2006; Carreño et al., 2007a,b; Birkmann and Fernando, 2008; Barbat et al., 2011; Carreño et al., 2012). A core element of these approaches is a feedback-loop system which underlines that vulnerability is dynamic and that vulnerability assessment cannot be limited to the identification of deficiencies, but rather also take into account the potential feedback loops and intervention tools that exist or can be developed in order to reduce vulnerability. Moreover, the approaches of so called integrative and holistic frameworks also incorporate the perspective of sustainable development into the assessment of vulnerability (Birkmann, 2006b).

The fourth school of thought emerged within the context of CCA (see e.g., Füssel and Klein, 2006). Most of these approaches focus closely on the definition of vulnerability used by the IPCC (AR4). Vulnerability in this regard is understood as a function of exposure, sensitivity,[3] and adaptive capacities (Füssel, 2007a,b; McCarthy et al., 2001; IPCC, 2007; O'Brien et al., 2008a,b). These frameworks, however, differ from the understanding of vulnerability in the DRM community in that they take into account the rate and magnitude of climate change. This introduces a critical distinction between the understanding of vulnerability within climate change and the other schools of thought discussed above. The concept of vulnerability here includes external environmental factors of shock or stress. Hence, in this view, the magnitude and frequency of potentially hazardous events is to be included in the calculation of vulnerability to climate change and hence, the vulnerability concept shifts toward a risk definition.

1.3 MULTIDIMENSIONAL AND HOLISTIC PERSPECTIVE: THE MOVE FRAMEWORK

1.3.1 Goals of the Framework

A key goal when developing the MOVE framework was to provide an improved conceptualization of the multifaceted nature of vulnerability, accounting for key causal factors such as exposure, susceptibility, lack of resilience (lack of societal response capacities), as well as for the different thematic dimensions of vulnerability: physical, social, ecological, economic, cultural, and institutional.

3. Interestingly, the term sensitivity means different things to different communities; however, the actual factors used to assess sensitivity of a system in CCA can be closely linked to factors that are used to characterize susceptibility or fragility in the DRM context.

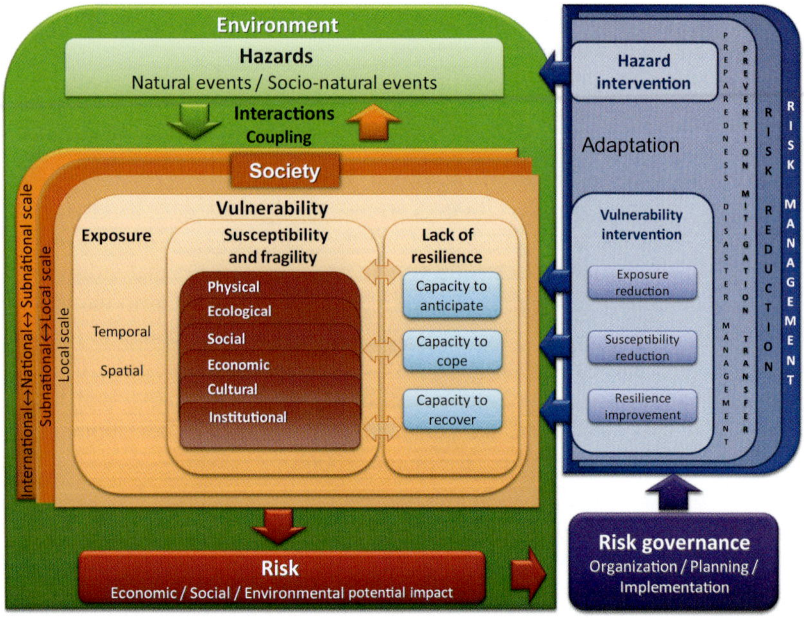

FIGURE 1.1 The MOVE framework. *Own figure, based particularly on concepts of Cardona, 1999a, 2001, p. 65; Turner et al., 2003; Bogardi and Birkmann, 2004; IDEA, 2005; Birkmann, 2006b; Carreño et al., 2007a; see also Birkmann et al., 2013.*

Additionally, the framework incorporates the concept of adaptation into DRM, and therewith explicitly differentiates coping from adaptation.

The MOVE conceptual framework (see Figure 1.1) underlines that hazards are of natural or socio-natural origin while vulnerability in its multifaceted nature is mainly linked to societal conditions and processes.

1.3.2 Key Factors of Vulnerability within the MOVE Framework

At its core, the MOVE framework differentiates key factors of vulnerability and shows the different thematic dimensions of vulnerability. The key factors of vulnerability are defined as follows:

1. Exposure describes the extent to which a unit of assessment falls within the geographical range of a hazard event. Exposure extends to fixed physical attributes of social systems (infrastructure) but also human systems (livelihoods, economies, cultures) that are spatially bound to specific resources and practices that may also be exposed. Exposure is then qualified in terms of spatial and temporal patterns.
2. Susceptibility (or fragility) describes the predisposition of elements at risk (social and ecological) to suffer harm. Although susceptibility and fragility imply subtle differences in various concepts, we mainly use them synonymously

within the *meta*-framework in order to emphasize the core differences between exposure, susceptibility, and lack of resilience. In this context, susceptibility (or fragility) can be calculated and addressed often independent of exposure.

3. Lack of resilience or societal response capacity is determined by limitations in terms of access to and mobilization of the resources of a community or a social-ecological system in responding to an identified hazard. This includes preevent risk reduction, in-time coping, and postevent response measures. Compared to adaptation processes and adaptive capacities, these capacities focus mainly on the ability to maintain the system in light of a hazard event impacting the system or element exposed. In this sense, the capacity to anticipate, the capacity to cope, and the capacity to recover can include significant changes to existing practices around a referent hazard event/scenario but does not include learning based on the potential for future change in hazard and vulnerability contexts. However, the concept of resilience also includes learning and reorganization processes, and therefore is positioned as a subcomponent of the adaptation box. Compared to the key factor "lack of resilience", which refers to existing capacities, the adaptation box also deals with the ability of a community or a system to learn from past disasters and to change existing practices for potential future changes in hazards as well as vulnerability contexts.
4. Hazard is used to describe the potential occurrence of natural, socio-natural, or anthropogenic events that may have physical, social, economic, and environmental impact in a given area and over a period of time. Therefore, hazard is defined by the potentiality of geodynamics or hydro-meteorological processes to cause effects upon exposed elements. In addition, the concept of coupling emphasizes the framework's assertion that any defined hazard is given form and meaning by interaction with social systems, and similarly, social systems are influenced by their actual and perceived hazard context.

1.3.2.1 Multidimensional Vulnerability

In addition to key factors of vulnerability, core thematic dimensions of vulner ability have to be addressed within a holistic assessment process. Key thematic components are explained as follows:

- *Social dimension*: propensity for human well-being to be damaged by disruption to individual (mental and physical health) and collective (health, education services, etc.) social systems and their characteristics (e.g., gender, marginalization of social groups).
- *Economic dimension*: propensity for loss of economic value from damage to physical assets and/or disruption of productive capacity.
- *Physical dimension*: potential for damage to physical assets including built-up areas, infrastructure, and open spaces.
- *Cultural dimension*: potential for damage to intangible values including meanings placed on artifacts, customs, habitual practices, and natural or urban landscapes.

- *Environmental dimension*: potential for damage to all ecological and biophysical systems and their different functions. This includes particular ecosystem functions and environmental services (see e.g., Renaud, 2006) but excludes cultural values that might be attributed.
- *Institutional vulnerability*: potential for damage to governance systems, organizational form and function, as well as guiding formal/legal and informal/customary rules—any of which may be forced to change following weaknesses exposed by disaster and response.

The majority of assets and systems exposed to hazard will exhibit more than one dimension of vulnerability.

1.3.2.2 Risk and Risk Governance

In contrast to vulnerability, risk is defined as the probability of harmful consequences or losses resulting from interactions between hazard and vulnerable conditions. It is the potential for physical, social, economic, environmental, cultural, or institutional consequences or losses, in a given area and over a period of time (see UNISDR, 2004). In addition, risk governance is linked to decisions and actions performed by formal stakeholders such as governments or different governmental institutions and informal stakeholders (individual households) that include tasks on risk reduction, prevention, mitigation, and transfer, and also preparedness and disaster management (see e.g., Renn, 2008).

1.3.2.3 Adaptation

Adaptation and adaptive capacities describe techniques, assets, and strategies applied or available to intervene in vulnerability; i.e., manage exposure, susceptibility, and resilience at any one moment in time. Compared to the concept of coping, adaptation here is understood as a mechanism and response capacity that also aims to change existing institutional and organizational structures and hence does intend to modify the system; while coping is associated with capacities to maintain a system as it is in the face of adverse consequences. In this regard, resilience building and improvement is seen as a component of adaptation. Compared to capacities to cope or to recover (classified as the "lack of resilience"), resilience in the adaptation box refers to learning and reorganization processes in the light of potential future changes and hazards, as well as potential changes in vulnerability. Consequently, the concept of resilience is differentiated into (1) a more reactive part that refers to the lack of resilience, while (2) the ability to learn and to reorganize in anticipation of future changes (proactive actions) are linked to the notion of adaptation. While coping capacities and resilience are primarily linked to capacities that help to maintain the current status of the systems under stress, adaptation as a concept implies actions aimed at making more profound change in socio-ecological relations (see e.g., Pelling, 2010; Birkmann, 2010, 2013).

1.3.3 Theoretical Grounding of the Concept: System Thinking and Nonlinearity

In terms of the underlying theoretical concept, the framework is based on and refers to general systems theory, cybernetics and interlinked systems theory (see e.g., Vester, 2008). It underlines that vulnerability and risk are embedded in processes and therefore have a dynamic nature—they change over time. Although this has been stressed by different authors in the past (see e.g., Keiler et al., 2006; Bründl et al., 2010), many vulnerability and risk assessment approaches in the context of natural hazards and climate change still do not account for capacities to cope and to adapt and hence do not link their thinking to newer approaches in the context of disaster risk reduction and CCA. Compared with the linear understanding of feedback and response processes held by cybernetic theory and the idea that systems can be steered more or less easily, the MOVE framework is based on the understanding that risks also involve complexity and the emergence of different vulnerability as well as response patterns to risks observed. This thinking can build on a legacy of engagement with cybernetics in the geographical tradition of hazards studies (Cardona, 2001, 2010; Hilhorst, 2004; Pelling, 2010). Critics argued that while the cybernetic approach had made progress in providing a framework that recognized social context as a mediating pressure on the environment, it did not have the conceptual tools to analyze these relationships. Deeper social relations of production and power were not included.

Current perspectives on adaptation, vulnerability, and risk to climate change developed from systems thinking within the socio-ecological systems school have made some advances in integrating power dynamics into models of risk (see e.g., Gunderson and Holling, 2002). The policy warning is that social systems will change in time driven either by environmental crisis or preemptive risk management (Handmer and Dovers, 1996). It is possible to insert power into this analysis through the interaction of institutions or structures and agency in cycles of adaptation. The challenge here is to understand under which conditions institutions and decision makers are able to deal with interacting social and ecological crises (Galaz et al., 2011).

1.4 THE APPLICATION: CRITERIA AND INDICATORS

The MOVE framework is first and foremost a thinking tool, however, it also served as an important guiding vision for the development of specific criteria and indicators that allow to assess vulnerability in different cities and regions in Europe and worldwide. Within the application of the framework and the development of concrete assessment tools, the constraints and limits of the framework become evident. However, interestingly, various case studies presented in this volume show that many case studies could apply the framework within the empirical research process or selected components of it.

The case studies dealing with seismic risk in Barcelone (Spain) and the flood vulnerability assessment in the Salzach River Basin as well as the floods in Cologne use various core factors of vulnerability outlined in the MOVE framework within their assessment. In this regard, important indicators for assessing (1) exposure, (2) susceptibility, and (3) response capacities in terms of the lack of resilience (lack of coping or recovery capacities) as well as adaptation processes are shown. Consequently, the framework can be operationalized for specific assessments. More specific vulnerability and risk assessments are conducted within the context of the healthcare system in Florence (Italy) and the heatwave vulnerability and risk management capacities in London (UK). Vulnerability assessments to forest fires and coastal erosion is demonstrated within the case study of Portugal and a vulnerability assessment to mass movements in mountainous environments is show with the case study of South Tyrol.

Overall, the examples show that most of the components of the MOVE framework can be applied as a basis for developing and differentiating indicators and criteria for vulnerability assessment within the broader context of risk management and risk governance. However, the interested reader will also find important differences in the operationalization and further concretization of different vulnerability factors and concepts such as adaptation and coping. These differences could be seen as a cacophony of the concept and term again; however, we view these differences and specific perspectives also as an important part of the contextualization of the general MOVE framework and the concept of vulnerability. That means compared to a standardized natural hazard assessment, for example, in terms of earthquake hazards using the Richter or Moment Magnitude scale, vulnerability about societal conditions and hence has to be translated and operationalized within a specific context. Consequently, some indicators and core factors of vulnerability might need to be interpreted differently or have to include additional information and indicators in order to make sense for the specific spatial, cultural, or socio-economic context. The core message that the MOVE framework provides for DRM and CCA is that even if the exposure of people or infrastructures or ecosystems is similar, the vulnerability of different groups, infrastructures, and ecosystems is most likely to be different, since vulnerability is differential. For example, in the case of the assessment of social vulnerability in Barcelona, low-income groups and low-income urban areas were used as proxies to assess particularly vulnerable groups and areas. The lack of institutional resilience was assessed by using expert interviews and a benchmark index for preparedness and the capacity of different agencies to deal with earthquake-related disasters (rescue teams, fire brigades, etc.) in Barcelona. Hence, the assessment of response capacities requires a more precise understanding of the hazard and specific vulnerabilities that should be considered when evaluating response and preparedness strategies and capacities. Overall, the case studies provide a rich overview of different vulnerability facets and the approach to systematize important information according to the framework outlined before.

1.5 CHALLENGES AND OUTLOOK

Against the background of the diverse approaches and assessment methods used in DRM and CCA, the MOVE framework presented and the further application of it in case study cities or regions (see case study chapters) shows that a common *meta*-approach can be achieved linking different approaches and research communities. In this regard it is also important to note that the MOVE framework and its discussion also informed particularly the new framing of risk management within the Intergovernmental Panel on Climate Change (IPCC). The IPCC Special report on "Managing the Risk of Extreme Events to Advance Climate Change Adaptation" is based on this broader framing of vulnerability and the understanding that extreme events do not necessarily cause extreme impacts. Rather the new framework of the IPCC underscores that next to the physical phenomena (e.g., hazard or extreme weather event), exposure and vulnerability are key in determining whether such extreme events can cause extreme impacts (see IPCC, 2012).

As a heuristic, the MOVE framework is a thinking tool to guide systemic assessment processes and the development of indicators, which for example could be described in specific guiding documents (Vinchon et al., 2011). However, the framework does not provide a specific assessment method (qualitative or quantitative) or a predefined list of indicators. Rather, it outlines key factors and different dimensions of vulnerability that can serve as a basis for a systematic operationalization of vulnerability (see the following chapters). The framework stresses the fact that many interactions that shape vulnerability are dynamic (change over time) and characterized by nonlinearity and place-specific factors. Thus, the application of the framework has to consider the place-specific characteristics that influence vulnerability and its components as well as the coupling processes between social and environmental systems. In addition, this framework considers different scales: not only recognizing the fact that certain characteristics are typical or only valid at a certain scales (e.g., community versus global assessment) but also reflecting the fact that specific scales (spatial and temporal scales) correspond with different needs of stakeholders and institutions operating at different times and spatial ranges. The specific translation of the framework into concrete measures also depends on the research object or subject (social group, physical buildings, socio-ecological systems) and the hazard context. An important benefit of the framework can be seen in the ability to straddle multiple approaches and epistemologies in natural and social sciences and DRM. Instead of focusing solely on the deficiencies of a community or incapacities of different social groups or social-ecological systems, the concept and its application shows that vulnerable groups or systems have also developed capacities that help them to survive or deal with changing environmental conditions.

Lastly, the framework is easy to understand for different disciplines and therefore enables and promotes the communication process between different communities, particularly between the DRM and CCA community. In this context, the framework also has a strong relevance for policy makers that aim to base their decisions on a comprehensive and integrative approach of vulnerability and risk identification. For example, the EU Flood Directive on Assessment and Management of Flood Risks (2007/60/EC) and the subsequent EU Flood Directive Implementation Strategy lack to address vulnerability in a comprehensive and integrative approach. In this regard, the framework could be used to inform agencies responsible for carrying out flood risk maps in the context of the directive, in terms of providing them with a broader overview of the various dimensions and key factors that should be considered within a holistic approach.

However, challenges remain with regard to the implementation of the framework and its key components in highly diverse hazards and cultural context situations. In particular, the intangible factors that determine institutional or cultural vulnerability are difficult to capture and to assess. However, the selected examples shown in the table provide an illustration on how one can capture these rather intangible aspects within such assessments. Hence, it is proven that these facets of vulnerability can be translated into assessable criteria and indicators.

REFERENCES

Adger, W.N., 2006. Vulnerability. Global Environmental Change 16 (3), 268–281.

Alexander, D.E., 1993. Natural Disasters. UCL Press Limited, London.

Alexander, D.E., 2000. Confronting Catastrophe. Terra Publishing, Harpenden.

Aysan, Y., 1993. Vulnerability assessment. In: Merriman, P.A., Browitt, C.W.A. (Eds.), Natural Disasters: Protecting Vulnerable Communities(London).

Barbat, A.H., Carreño, M.L., Cardona, O.D., Marulanda, M.C., 2011. Evaluación holística del riesgo sísmico en zonas urbanas. Revista Internacional de Métodos Numéricos para Cálculo y Diseño en Ingeniería 27, 3–27.

Berkes, F., Colding, J., Folke, C., 2003. Introduction. In: Berkes, F., Colding, J., Folke, C. (Eds.), Navigating Social-Ecological Systems: Building Resilience for Complexity and Change. Cambridge University Press, Boston? London?

Birkmann, J. (Ed.), 2013. Measuring Vulnerability to Natural Hazards - Towards Disaster Resilient Societies. , Second ed.United Nations University Press, Tokyo, New York, Paris.

Birkmann, J. (Ed.), 2006a. Measuring Vulnerability to Natural Hazards - Towards Disaster Resilient Societies. United Nations University Press, Tokyo, New York, Paris.

Birkmann, J., 2006b. Measuring vulnerability to promote disaster-resilient societies: conceptual frameworks and definitions. In: Birkmann, J. (Ed.), Measuring Vulnerability to Natural Hazards: Towards Disaster Resilient Societies. United Nations University Press, Tokyo, New York, Paris, pp. 9–54.

Birkmann, J., 2006c. Indicators and criteria for measuring vulnerability: theoretical bases and requirements. In: Birkmann, J. (Ed.), Measuring Vulnerability to Natural Hazards - Towards Disaster Resilient Societies. United Nations University Press, Tokyo, New York, Paris, pp. 55–77.

Birkmann, J., 2010. Regulation and coupling of society and nature in the context of natural hazards – different theoretical approaches and conceptual frameworks and their applicability to analyse social-ecological crises phenomena. In: Brauch, H.G., Oswald Spring, U., Mesjasz, C., Grin, J., Kameri-Mbote, P., Chourou, B., Dunay, P., Birkmann, J. (Eds.), Coping with Global Environmental Change, Disasters and Security - Threats, Challenges, Vulnerabilities and Risks. Hexagon Series on Human and Environmental Security and Peace, vol. 5. Springer, Berlin, Heidelberg, New York.

Birkmann, J., Buckle, P., Jaeger, J., Pelling, M., Setiadi, N., Garschagen, M., Fernando, N., Kropp, J., 2010. Extreme events and disasters: a window of opportunity for change? Analysis of changes, formal and informal responses after mega-disasters. Natural Hazards 55 (3), 637–655.

Birkmann, J., Cardona, O.D., Carreno, L., Barbat, A., Pelling, M., Schneiderbauer, S., Kienberger, S., Kelier, M., Alexander, D., Zeil, P., Welle, T., 2013. Framing vulnerability, risk and societal responses: the MOVE framework. Natural Hazards 67 (2), 193–211.

Birkmann, J., Fernando, N., 2008. Measuring revealed and emergent vulnerabilities of coastal communities to Tsunamis in Sri Lanka. Disasters 32 (1), 82–104.

Birkmann, J., von Teichman, K., 2010. Integrating disaster risk reduction and climate change adaptation: key challenges—scales, knowledge, and norms. Sustainability Science 5 (2), 171–184.

Blaikie, P., Cannon, T., Davis, I., Wisner, B., 1994. At Risk: Natural Hazards, People, Vulnerability, and Disasters. Routledge, London.

Blaikie, P., Cannon, T., Davis, I., Wisner, B., 1996. Vulnerabilidad, el Entorno Social de los Desastres. La RED-ITDG, Bogota, D.C.

Bogardi, J., Birkmann, J., 2004. Vulnerability assessment: the first step towards sustainable risk reduction. In: Malzahn, D., Plapp, T. (Eds.), Disasters and Society – from Hazard Assessment to Risk Reduction. Logos, Berlin, pp. 75–82.

Bohle, H.-G., 2008. Krisen, Katastrophen, Kollaps – Geographien von Verwundbarkeit in der Risikogesellschaft. In: Kulke, E., Popp, H. (Eds.), Umgang mit Risiken. Katastrophen – Destabilisierung – Sicherheit. Lausitzer Druck- und Verlagshaus GmbH, Bayreuth, pp. 69–82. Deutscher Geographentag 2007, Bautzen.

Bohle, H.-G., Glade, T., 2008. Vulnerabilitätskonzepte in Sozial- und Naturwissenschaften. In: Felgentreff, C., Glade, T. (Eds.), Naturrisiken und Sozialkatastrophen. Spektrum Verlag, Berlin, Heidelberg, pp. 99–119.

Boin, A., McConnell, A., 2007. Preparing for critical infrastructure breakdowns: the limits of crisis management and the need for resilience. Journal of Contingencies and Crisis Management 15 (1), 50–59.

Bonanno, G.A., Galea, S., Bucciarelli, A., Vlahov, D., 2006. Psychological resilience after disaster: New York City in the aftermath of the September 11th terrorist attack. Psychological Science 17 (3), 181–186.

Bonanno, G., 2008. Loss, trauma, and human resilience: have we underestimated the human capacity to thrive after extremely aversive events? American Psychologis 59 (1), 20–28.

Bründl, M., Bartelt, P., Schweizer, J., Keiler, M., Glade, T., 2010. Snow avalanche risk analysis - review and future challenges. In: Alcantara-Ayla, I., Goudie, A. (Eds.), Geomorphological Hazards and Disaster. Cambridge University Press, Boston, pp. 49–61.

Burton, I., Kates, R.W., White, G.F., 1993. The Environment as Hazard. Guildford Press, London

Cannon, T., 1994. Vulnerability analysis and the explanation of 'Natural' disasters. In: Varley, A. (Ed.), Disasters, Development and Environment. John Wiley and Sons, Chichester, pp. 13–29.

Cannon, T., 2006. Vulnerability analysis, livelihoods and disasters. In: Ammann, W.J., Dannenmann, S., Vulliet, L. (Eds.), Risk 21: Coping with Risks Due to Natural Hazards in the 21st Century. Taylor and Francis Group plc, London, pp. 41–49.

Cardona, O.D., 1986. Estudios de Vulnerabilidad y Evaluación del Riesgo Sísmico: Planificación Física y Urbana en Áreas Propensas. Boletín Técnico de la Asociación Colombiana de Ingeniería Sismica 33 (2), 32–65.

Cardona, O.D., 1990. Terminología de Uso Común en Manejo de Riesgos. AGID Reporte No. 13, EAFIT. In: Fernández, M.A. (Ed.), Medellín, actualizado y reimpreso en ciudades en riesgo. La RED, USAID.

Cardona, O.D., 1999a. Diagnóstico local de riesgos naturales en Santa Fe de Bogotá para la planificación y medidas de mitigación. Panamericana-Secretaría de Salud, Bogotá, D.C.

Cardona, O.D., 1999b. Environmental management and disaster prevention: two related topics. In: Fernandez, M.A. (Ed.), Cities at Risk: Environmental Degradation, Urban Risks and Disasters in Latin America. A/H Editorial, La RED, US AID, Quito, pp. 77–102.

Cardona, O.D., 2001. Estimación Holística del Riesgo Sísmico utilizando Sistemas Dinámicos Complejos.(Doctoral dissertation) Technical University of Catalonia, Department of Terrain Engineering. Available from: http://www.desenredando.org/public/varios/2001/ehrisusd/index.html.

Cardona, O.D., 2004. The need for rethinking the concepts of vulnerability and risk from a holistic perspective: a necessary review and criticism for effective risk management. In: Bankoff, G., Frerks, G., Hilhorst, D. (Eds.), Mapping Vulnerability: Disasters, Development and People. Publishers, London, pp. 37–51.

Cardona, O.D., 2011. Disaster risk and vulnerability: notions and measurement of human and environmental insecurity. In: Brauch, H.G., Oswald Spring, U., Mesjasz, C., Grin, J., Kameri-Mbote, P., Chourou, B., Dunay, P., Birkmann, J. (Eds.), Coping with Global Environmental Change, Disasters and Security - Threats, Challenges, Vulnerabilities and Risks. Hexagon Series on Human and Environmental Security and Peace Vol. 5. Springer, Berlin, Heidelberg, New York, pp. 107-121.

Carreño, M.L., 2006. Técnicas innovadoras para la evaluación del riesgo sísmico y su gestión en centros urbanos: Acciones ex ante y ex post(Doctoral dissertation) Technical University of Catalonia, Department of Terrain Engineering. Available from: http://www.tdx.cat/TDX-1102106-110455.

Carreño, M.L., Cardona, O.D., Barbat, A.H., 2007a. Urban seismic risk evaluation: a holistic approach. Natural Hazards 40 (1), 137–172.

Carreño, M.L., Cardona, O.D., Barbat, A.H., 2007b. Disaster risk management performance index. Natural Hazards 41 (1), 1–20.

Carreño, M.L., Cardona, O.D., Barbat, A.H., 2012. New methodology for urban seismic risk assessment from a holistic perspective. Bulletin of Earthquake Engineering 10 (2), 547–565.

Chambers, Robert, 1989. Vulnerability, coping and policy. Institute of Development Studies Bulletin 20 (2), 1–7.

Cutter, S., Barnes, L., Berry, M., Burton, C., Evans, E., Tate, E., Webb, J., 2008. A place-based model for understanding community resilience to natural disasters. Global Environmental Change 18 (4), 598–606.

Cutter, S., Finch, C., 2008. Temporal and spatial changes in social vulnerability to natural hazards. PNAS 105 (7), 2301–2306.

Cutter, S.L., Boruff, B.J., Shirley, W.L., 2003. Social vulnerability to environmental hazards. Social Science Quarterly 84 (2003), 242–261.

DFID (Department for International Development), 1999. Sustainable Livelihood Guidance Sheets. (London).

Folke, C., 2006. Resilience: the emergence of a perspective for social–ecological systems analyses. Global Environmental Change 16, 253–267.

Fuchs, S., 2009. Susceptibility versus resilience to mountain hazards in Austria – paradigms of vulnerability revisited. Natural Hazards and the Earth System Sciences 9, 337–352.

Füssel, H.M., 2007a. Adaptation planning for climate change: concepts, assessment approaches and key lessons. Sustainability Science 2 (2), 265–275.

Füssel, H.M., 2007b. Vulnerability: a generally applicable conceptual framework for climate change research. Global Environmental Change 17 (2), 155–167.

Füssel, H.M., Klein, R., 2006. Climate change vulnerability assessment: an evolution of conceptual thinking. Climatic Change 75 (3), 301–329.

Galaz, V., Moberg, F., Olsson, E.-K., Paglia, E., Parker, C., 2011. Institutional and political leadership dimensions of cascading ecological crises. Public Administration 89 (2), 361–380.

GAR (Global Assessment Report), 2011. Revealing Risk, Redefining Development. In: UN/ISDR (Ed.), Geneva.

Gunderson, L.H., Holling, C.S. (Eds.), 2002. Panarchy: Understanding Transformations in Social-ecological Systems. Island Press, London.

Handmer, J.W., Dovers, S.R., 1996. A typology of resilience: rethinking institutions for sustainable development. Organization and Environment 9 (4), 482–511.

Hewitt, K. (Ed.), 1983. Interpretations of Calamity. Allen and Unwin, London.

Hilhorst, D., 2004. Complexity and diversity: unlocking social domains of disaster response. In: Bankoff, G., Frerks, G., Hilhorst, D. (Eds.), Mapping Vulnerability: Disasters, Development and People. Earthscan Publishers, London, pp. 52–67.

Holling, C., 1973. Resilience and stability of ecological systems. Annual Review of Ecology and Systematics 4, 1–23.

IDEA, 2005. Indicators of Disaster Risk and Risk Management – Main Technical Report, English and Spanish edition. National University of Colombia/Manizales, Institute of Environmental Studies/IDEA, Inter-American Development Bank, Washington, D.C. Available from:. http://idea.unalmzl.edu.co.

IPCC, 2007. In: Parry, M.L., Canziani, O.F., Palutikof, J.P., van der Linden, P.J., Hanson, C.E. (Eds.), Climate Change 2007: Impacts, Adaptation and Vulnerability, Contribution of Working Group II to the Fourth Assessment Report of the Intergovernmental Panel on Climate Change. Cambridge University Press, Cambridge, pp. 7–22.

IPCC, 2012. Managing the Risk of Extreme Events and Disasters to Advance Climate Change Adaptation. Cambridge University Press, Cambridge. Special Report of the Intergovernmental Panel on Climate Change (IPCC).

IPCC, 2013. In: Stocker, T.F., Qin, D., Plattner, G.-K., Tignor, M., Allen, S.K., Boschung, J., Nauels, A., Xia, Y., Bex, V., Midgley, P.M. (Eds.), Climate Change 2013: The Physical Science Basis. Working Group I Contribution to the Fifth Assessment Report of the Intergovernmental Panel on Climate Change. Summary for Policymakers. Cambridge University Press, Cambridge, United Kingdom and New York, NY, USA.

IPCC, 2014. In: Field, C.B., Barros, V.R., Mastrandrea, M.D., Mach, K.J., Climate Change 2014: Impacts, Adaptation, and Vulnerability. Working Group II Contribution to the Fifth Assessment Report of the Intergovernmental Panel on Climate Change. Summary for Policymakers. Cambridge University Press, Cambridge, United Kingdom and New York, NY, USA.

Keiler, M., Knight, J., Harrison, S., 2010. Climate change and geomorphological hazards in the Eastern European Alps. Philosophical Transactions of the Royal Society A 368, 2461–2479.

Keiler, M., Sailer, R., Jörg, P., Weber, C., Fuchs, S., Zischg, A., Sauermoser, S., 2006. Avalanche risk assessment - a multi-temporal approach, results from Galtür, Austria. Natural Hazards and the Earth System Sciences 6, 637–651.

Kelly, P.M., Adger, W.N., 2000. Theory and practice in assessing vulnerability to climate change and facilitating adaptation. Climatic Change 47 (4), 325–352.

Kienberger, S., Lang, S., Zeil, P., 2009. Spatial vulnerability units – expert-based spatial modelling of socio-economic vulnerability in the Salzach catchment, Austria. Nature Hazards the Earth System Sciences 9, 767–778.

Lavell, A., 1999. Environmental degradation, risks and urban disasters. Issues and concepts: towards the definition of a research agenda. In: Fernandez, M.A. (Ed.), Cities at Risk: Environmental Degradation, Urban Risks and Disasters in Latin America. A/H Editorial, La RED, US AID, Quito, pp. 19–58.

Liverman, D.M., 1990. Chapter 26-vulnerability to global environmental change. In: Kasperson, R.E., Dow, K., Golding, D., Kasperson, J.X. (Eds.), Understanding Global Environmental Change: The Contributions of Risk Analysis and Management. Clark University, Worcester, MA, pp. 27–44.

Maskrey, A., 1984. Community based hazard mitigation. In: Disasters Mitigation Program Implementation. Virginia Polytechnic Institute.

McCarthy, J.J., Canziani, O.F., Leary, N.A., Dokken, D.J., White, K.S. (Eds.), 2001. Climate Change 2001: Impacts, Adaptation, and Vulnerability. Working Group II of the Intergovernmental Panel on Climate Change. Cambridge University Press, Cambridge.

Mitchell, T., van Aalst, M.K., 2009. Convergence of Disaster Risk Reduction and Climate Change Adaptation - A Review for DFID. IDS, Brighton.

O'Brien, K., Eriksen, S., Schjolen, A., Nygaard, L., 2004. What's in a Word? Conflicting Interpretations of Vulnerability in Climate Change Research. CICERO Working Paper, No. 4. CICERO, Oslo University.

O'Brien, K., Eriksen, S., Schjolen, A., Nygaard, L., 2007. Why different interpretations of vulnerability matter in climate change discourses. Climate Policy 7, 73–88.

O'Brien, K., O'Keefe, P., Meena, H., Rose, J., Wilson, L., 2008b. Climate adaptation from a poverty perspective. Climate Policy 8 (2), 194–201.

O'Brien, K., Sygna, L., Leinchenko, R., Adger, W.N., Barnett, J., Mitchell, T., Schipper, L., Tanner, T., Vogel, C., Mortreux, C., 2008a. Disaster risk reduction, climate change adaptation and human security. GECHS Report 3.

Oxfam America, 2009. Exposed – Social Vulnerability and Climate Change in the US. Southeast Press, Boston.

Papathoma-Köhle, M., Kappes, M.S., Keiler, M. & T., Glade, 2011. Physical vulnerability assessment for alpine hazards: state of the art and future needs future needs. Natural Hazards 58 (2), 645–680.

Pelling, Mark, 1997. What determines vulnerability to floods; a case study in Georgetown, Guyana. Environment and Urbanization 9 (1), 203–226.

Pelling, M., 2001. Natural disasters? In: Castree, N., Braun, B. (Eds.), Social Nature. Blackwells, London.

Pelling, M., 2010. Adaptation to Climate Change: From Resilience to Transformation. Routledge, London.

Pelling, M., Dill, K., 2010. Disaster politics: tipping points for change in the adaptation of sociopolitical regimes. Progress in Human Geography 34 (1), 21–23.

Phillips, B.D., Fordham, M., 2009. Introduction: chapter 1. In: Phillips, B.D., Thomas, D.S.K., Fothergill, A., Blinn-Pike, L. (Eds.), Social Vulnerability to Disasters. CRC Press, Boca Raton, FL.

Quarantelli, E.L., 1998. What Is a Disaster? Routledge, New York.

Renaud, F.G., 2006. Environmental components of vulnerability. In: Birkmann, J. (Ed.), Measuring Vulnerability to Natural Hazards: Towards Disaster Resilient Societies. United Nations University Press, pp. 117–127.

Renn, O., 2008. Risk Governance – Coping with Uncertainty in a Complex World. Earthscan, London.

Romieu, E., Welle, T., Schneiderbauer, S., Pelling, M., Vinchon, C., 2010. Vulnerability assessment within climate change and natural hazard contexts: revealing gaps and synergies through costal applications. Sustainability Science 5 (2), 159–170.

Thomalla, F., Downing, T., Spanger-Siegfried, E., Han, G., Rockström, J., 2006. Reducing hazard vulnerability: towards a common approach between disaster risk reduction and climate adaptation. Disasters 30 (1), 39–48.

Timmerman, P., 1981. Vulnerability, Resilience and the Collapse of Society. Environmental Monograph No. 1. Institute for Environmental Studies, University of Toronto.

Turner, B.L., Kasperson, R.E., Matson, P.A., McCarthy, J.J., Corell, R.W., Christensen, L., Eckley, N., Kasperson, J.X., Luers, A., Martello, M.L., Polsky, C.-, Pulsipher, A., Schiller, A., 2003. A framework for vulnerability analysis in sustainability science. PNAS (Proceedings of the National Academy of Sciences) 100 (14), 8074–8079.

UNDRO, 1980. Natural disasters and vulnerability analysis. Report of experts group meeting of July 9–12, 1979, Geneva.

UNISDR, 2004. Living with Risk. UNISDR, Geneva.

UNISDR, 2009. Terminology on Disaster Risk Reduction. Geneva. Available from: http://unisdr.org/eng/library/lib-terminology-eng.htm.

Vester, F., 2008. Die Kunst vernetzt zu denken – Ideen und Werkzeuge für einen neuen Umgang mit Komplexität. Deutscher Taschenbuch Verlag, München.

Vinchon, C., Carreño, M.-L., Contreras-Mojica, D.M., Kienberger, S., Schneiderbauer, S., Alexander, D., Barbat, A.H., Cardona, O.D., Decker, B., Eidsvig, U., Papathoma-Köhle, M., Miniati, R., Pratzler-Wanczura, S., Ulbrich, T., Vangelsten, B.V., Welle, T., 2011. Assessing vulnerability to natural hazards in Europe: from principles to practice. MOVE Deliverable No.4.2. A Manual on Concept, Methodology and Tools. Available from: http://www.move-fp7.eu/documents/MOVE_Manual.pdf.

Welle, T., Birkmann, J., Rhyner, J., Witting, M., Wolfertz, J., 2012. World risk index 2012: concept and results. In: Alliance Development Works (Ed.), World Risk Report, pp. 11–26. Berlin.

Wisner, B., 2006. Self-assessment of coping capacity: participatory, proactive, and qualitative engagement of communities in their own risk management. In: Birkmann, J. (Ed.), Measuring Vulnerability to Natural Hazards - Towards Disaster Resilient Societies. United Nations University Press, Tokyo, New York, Paris, pp. 316–328.

Wisner, B., Blaikie, P., Cannon, T., Davis, I., 2004. At Risk, Natural Hazards, People's Vulnerability and Disasters. Routledge, London, New York.

Yohe, G., Tol, R.S.J., 2002. Indicators for social and economic coping capacity: moving toward a working definition of adaptive capacity. Global Environmental Change 12 (1), 25–40.

Chapter 2

Holistic Evaluation of Seismic Risk in Barcelona

Martha Liliana Carreño*, Alex H. Barbat†, Omar D. Cardona‡ and Mabel C. Marulanda§

*Centre Internacional de Mètodes Numèrics en Enginyeria (CIMNE), Universitat Politècnica de Catalunya, Barcelona, Spain, †Universitat Politècnica de Catalunya, Centre Internacional de Mètodes Numèrics en Enginyeria (CIMNE), Barcelona, Spain, ‡Universidad Nacional de Colombia, Sede Manizales, Colombia, §Centre Internacional de Mètodes Numèrics en Enginyeria (CIMNE) Barcelona, Spain

Chapter Outline

2.1 INTRODUCTION

This chapter shows the case study of Barcelona; here the seismic risk of the city is studied by applying a probabilistic analysis of the hazard and risk. On this basis a holistic evaluation of risk is performed taking into account social fragility and lack of resilience aspects. Finally, a complementary evaluation of the disaster risk management performance is conducted.

2.2 DESCRIPTION OF THE CITY

The city of Barcelona was founded by the Romans, as a result of the campaigns against the Carthaginian general Hannibal Barca during the Punic Wars. At the end of the Roman period, the city had nearly 12,000 inhabitants. By

Assessment of Vulnerability to Natural Hazards. http://dx.doi.org/10.1016/B978-0-12-410528-7.00002-3

the end of the fourth century, Barcelona was a fortified, walled town, covering about 10.5 Ha. Barcelona's evolution into a big city began in 1868 when adjacent, independent towns were incorporated as city districts. Between 1910 and 1930, the population grew from 587,411 to 1005,565 inhabitants. This population explosion was accompanied by a highly productive construction period. The city of Barcelona, capital of Catalonia and second largest city of Spain, has a total of 1621,537 inhabitants (2009) and is located on the northeast coast of Spain (Figure 2.1). Bounded by the Collserola ridge and rivers Besós and Llobregat, the city has an area of almost 100 km^2 and a population density over 16,215 persons/km^2. Barcelona is a leading tourist, economic, trade fair/exhibitions and cultural-sports center in Spain.

Nowadays, Barcelona has 10 administrative districts: Ciutat Vella, Eixample, Sants-Montjuïc, Les Corts, Sarrià-Sant Gervasi, Gràcia, Horta-Guinardó, Nou Barris, Sant Andreu, and Sant Martí. This division of the city is based on the historic evolution of the city. Ciutat Vella is the old center of the city and the Eixample is where the city expanded after the city walls were knocked down. The other districts correspond to municipal areas which appeared around the old city, outside the walls, and which became part of Barcelona during the nineteenth and twentieth centuries. The districts are subdivided into 73 neighborhoods and 235 AEBs (basic statistic areas in Spanish). According to the cadastral information,[1] there are 70,655 buildings in the city.

2.3 LOCAL SEISMIC HAZARD

Due to the high population growth, most part of the city's building stock was constructed when no seismic-resistant construction codes were required. The combination of very old buildings made of nonreinforced masonry and constructed without seismic requirements and a highly populated and active urban area can be extremely risky even under the effects of a moderate earthquake.

The seismicity of the Catalonia region is moderate compared to other regions along the Mediterranean Sea area. The fourteenth and fifteenth centuries were seismically active centuries; several earthquakes caused major damages in Barcelona (1373, 1410, 1427, 1428 and 1448). On February 2, 1428, an earthquake with the epicenter located in the Pyrenees with a local magnitude of 6.5 and an epicenter distance of 90 km damaged slightly some churches at Barcelona. This earthquake caused great damage to many houses and pulled down the rose window of Santa Maria de la Mar, and killed 16 people. In 1448, another earthquake with a local magnitude of 5.5 caused major damage to many properties both inside and outside the city. In the capital, it damaged Castell Nou producing a

1. Cadastal information was provided in 2009 by the Subdirecció d' Informació Cartogràfica i de Base, S.A. (ICB) (antes llamada Centre de Cartografia Automàtica, CCA), del Instituto Municipal de Informàtica del Ayuntamiento de Barcelona.

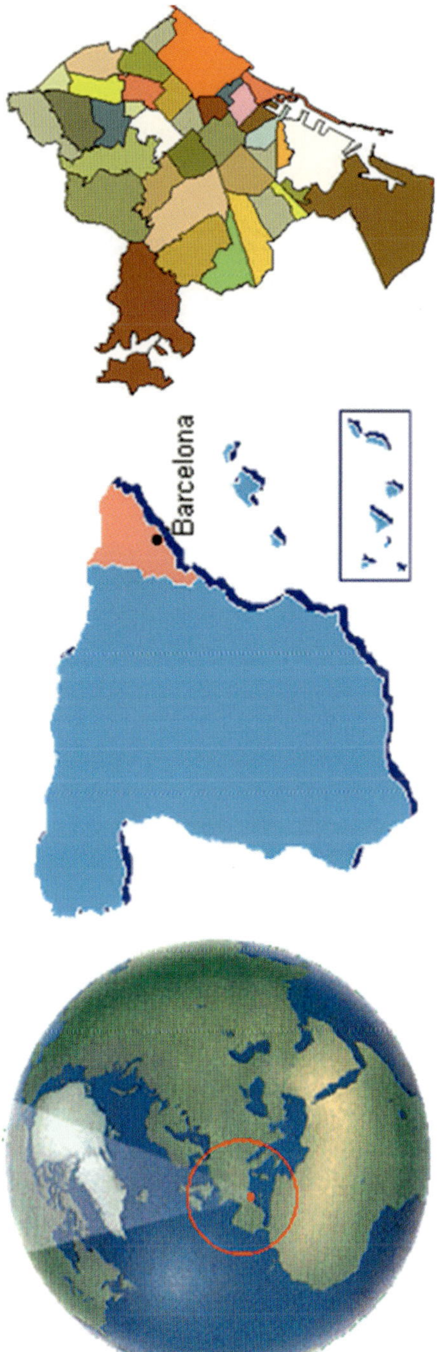

FIGURE 2.1 Location of the city of Barcelona.

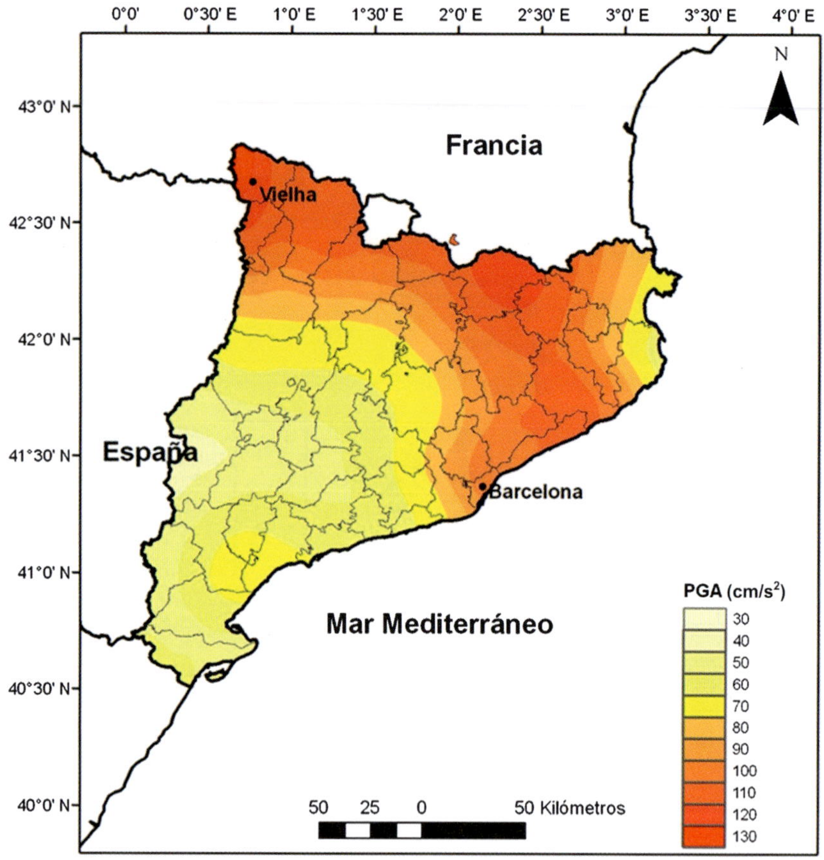

FIGURE 2.2 Seismic hazard map for 475 years of return period (GEOTER, 2008).

crack through which a man could easily pass. During the twentieth century, a few earthquakes had been felt in the city with a maximum intensity of IV degrees in the MSK intensity scale.

Various earthquakes struck Barcelona—and Catalonia—in the last third of the fourteenth century and the first half of the fifteenth century, which had a great effect on the collective memory compared to the actual damage (Egozcue et al., 1991). More recently, small earthquakes occurred (i.e., Mw:4.6 on May 15, 1995 and Ml: 4.0 on September 21, 2004), but without causing damage to people and buildings in the city. Figure 2.2 is a seismic hazard map of Catalonia which shows the peak ground accelerations (PGA) estimated for a 475 years return period. The seismic hazard has been simulated for Catalonia by using the system CRISIS 2007 (Ordaz, 2007) which is part of the platform ERN-CAPRA. CRISIS 2007 allows calculating the seismic hazard associated with all the feasible events, a group of selected events, or even a single relevant event, providing the probable maximum intensity for different occurrence rates or return period.

2.4 METHODOLOGY OF SEISMIC RISK ASSESSMENT

A probabilistic evaluation of the seismic risk has been done, including the probabilistic analysis of the seismic hazard which can affect Barcelona and of the seismic physical vulnerability of the buildings of the city. This evaluation involves the probabilistic calculation of both the seismic hazard and the vulnerability (Marulanda et al., 2013).

Parameters such as the frequency of occurrence of a given earthquake, the probability that it will occur at a specific site, the exceedance probabilities of the seismic intensities, etc., are included in the calculation models, in order to perform a probabilistic seismic hazard analysis (PSHA). The steps to be followed to apply PSHA are:

1. The characterization of the seismic sources generating earthquakes with influence on the study area, in terms of the geometry and the probability distribution of the points where the fault rupture starts. It is usual to assume a uniform probability distribution, which implies that the occurrence of an earthquake can be expected with the same probability in any point of the geometry of the defined source.
2. Determination of seismicity of the considered sources, based on historical records of events occurring in the defined geometry (seismic catalog), and on information and studies of neo-tectonics and paleo-seismology for the source. Seismicity is established through magnitude recurrence curves.
3. Selection of attenuation functions which will allow the hazard at the site to be completely characterized. Depending on the scope of the analysis, attenuation functions will be required for acceleration, velocity, displacement, spectral components of these parameters, duration, etc.
4. Finally, the uncertainties associated with location, size, and attenuation are combined, and a hazard code is obtained. This indicates the probability that a specific intensity may be equaled or exceeded in a given period of time.

The application to Barcelona takes into account the seismic sources identified by Secanell (2004) for the Catalonian region of Spain. Additionally, attenuation effects of the seismic waves were also considered in the mentioned reference by means of probabilistic spectral attenuation laws that include different source types and the local amplification effects based on microzonation studies (Ambraseys et al., 1996).

The seismic hazard is quantified in terms of return periods (or its inverse, the exceedance rates) of the relevant seismic intensities for the behavior of the structures. The exceedance rate of a seismic intensity is defined as the average number of times, per unit of time, in which the value of the seismic intensity is exceeded.

A series of stochastic events are simulated for the probabilistic risk analysis. These events represent the effects of all the possible earthquakes at any location and magnitude in the area of influence. The set of scenarios generated must represent all the hypocenter and the whole possible magnitudes associated with each hypocentral location. Each of these events or scenarios is associated with a specific occurrence frequency. The scenarios associated with earthquakes of minor magnitude have a greater probability of occurrence, while the scenarios associated with larger earthquakes have a relatively low probability of occurrence.

The probabilistic calculation method evaluates the desired risk parameters such as percentages of damage, economic losses, effects on people and other effects, for each of the hazard scenarios and then these results are probabilistically integrated by using the occurrence frequencies of each earthquake scenario. In Barcelona, 2058 seismic hazard scenarios have been generated by using the CAPRA platform.

Site effects are included to consider the amplification of seismic hazard parameters according to the geological characterization of Barcelona (Cid et al., 2001) (see Figure 2.3). Each zone is characterized by a transfer function and an amplification factor for the acceleration level on the rock. They are four zones: Zone I, Holocene deposits which include sand, silt, pebbles and organic matter; Zone II, Pleistocene deposits composed of clay, silt, gravel and calcareous crusts; Zone III, Paleozoic materials mainly granite, slate, limestone,

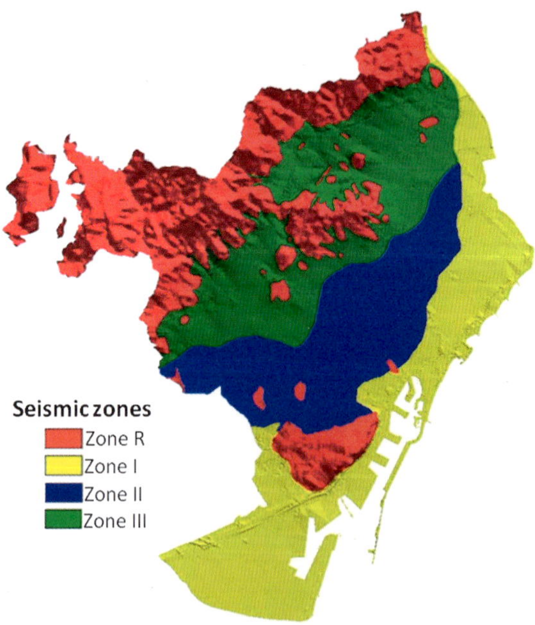

Soil zone	Vs (m/s)	Eurocode subsoil class
R	225	C
I	384	B
II	405	B
III	800	A

FIGURE 2.3 Seismic zonation based on local effects (Cid et al., 2001).

interbedded quartzite, sandstone and metamorphic rocks; and the Zone R corresponds to rock.

The *exposure* is mainly related to the infrastructure components or exposed population which can be affected by a particular event. To characterize the exposure, it is necessary to identify the individual components, including their location, main physical, geometric and engineering characteristics, their vulnerability to hazardous events, their economic value and the level of human occupation. The exposure value of the assets at risk is usually estimated from secondary sources such as available databases.

This study uses information of Barcelona compiled by Lantada (2007). The economic value of the exposed elements was supplied by the Cadastral Office of Barcelona in 2009, and they were 70,655 buildings taken into account. For each building, the geographic situation, economic value, year of construction, number of levels, structural type and human occupation were defined. Figure 2.4 shows the exposed value (in Euros) of the AEBs of Barcelona.

The *vulnerability* of the buildings in the city has been defined by vulnerability functions using the Vulnerability Module of the ERN-CAPRA platform starting from capacity curves defined in the previous studies performed in the RISK-UE project (ICC/CIMNE, 2004; Lantada et al., 2009, 2010; Irizarry et al., 2011). These functions are defined for each building typology; the most common structural system used in Barcelona is the unreinforced masonry, followed by the reinforced concrete, whose construction has increased rapidly in recent decades. Wood and steel structures are less used and these buildings are not usually for residential use but for industrial or other uses such as markets and sports areas among others. The used typologies were defined in ICC/CIMNE (2004) and are shown in Table 2.1, and were the masonry which is the most frequent building typology.

Each structural type is subdivided into three classes according to the height:

- Low height, L: one to two floors for masonry and wood structures; and one to three floors for reinforced concrete and steel buildings.
- Medium height, M: three to five floors for masonry and wood structures; and four to seven floors for reinforced concrete and steel buildings.
- High altitude, H: six or more floors for masonry and wood structures; and eight or more floors for reinforced concrete and steel buildings.

The vulnerability functions used are shown in Figure 2.5, for the unreinforced masonry buildings; and in Figure 2.6 for the rest of the building typologies. These vulnerability curves are relating spectral acceleration with mean damage ratio, but also include the standard deviation.

The physical seismic risk is evaluated by means of the convolution of the hazard with the vulnerability of the exposed elements and the results are the potential effects or consequences. Risk can be expressed in terms of damage or physical effects, absolute or relative economic loss and/or effects on the population. Once the expected physical damage has been estimated (average potential value and its

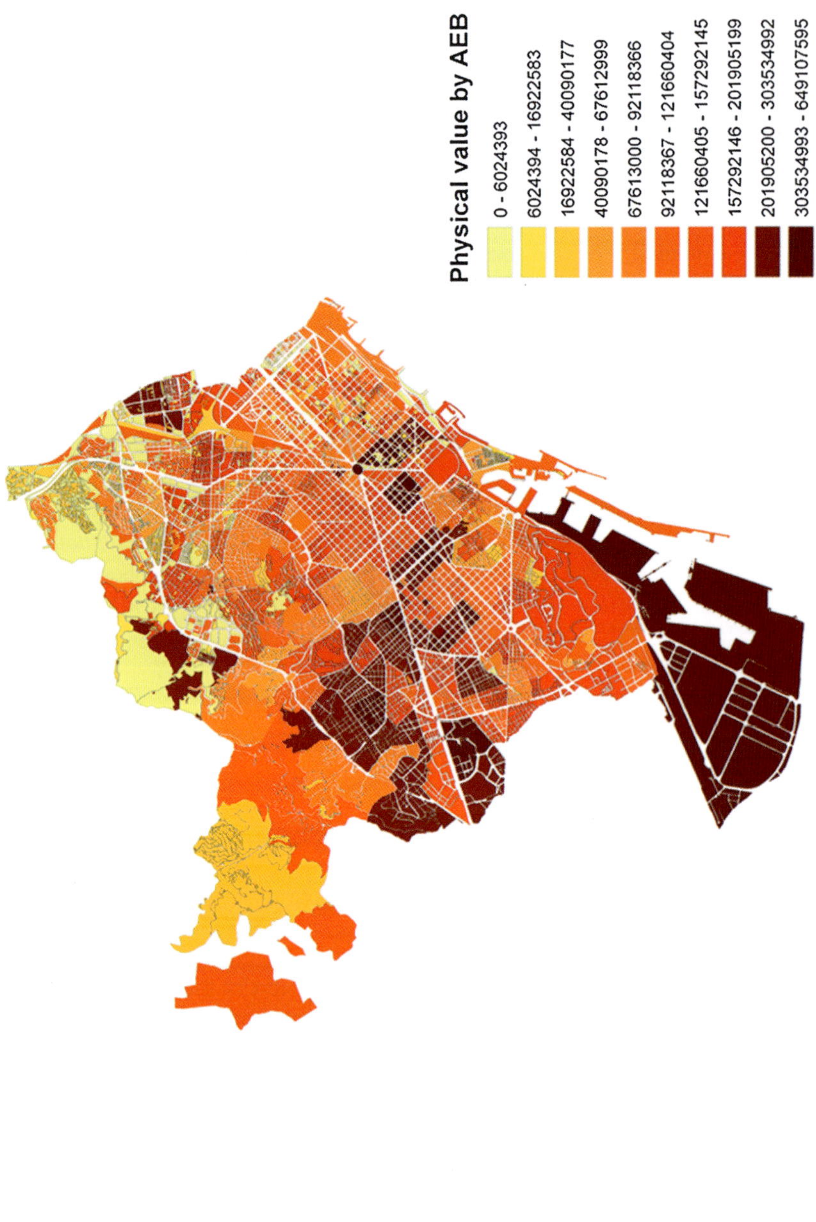

FIGURE 2.4 Exposed value of the AEBs of Barcelona.

TABLE 2.1 Building Typology Matrix for Barcelona (ICC/CIMNE, 2004)

Unreinforced masonry	M3.1	Unreinforced masonry bearing walls with wooden slabs
	M3.2	Unreinforced masonry bearing walls with Masonry vaults
	M3.3	Unreinforced masonry bearing walls with composite steel and masonry slabs
	M3.4	Reinforced concrete slabs
Reinforced concrete	RC3.1	Concrete frames with unreinforced masonry infill walls with regularly infill frames
	RC3.2	Concrete frames with unreinforced masonry infill walls with irregularly frames (i.e., irregular structural system, irregular infill, soft/weak storey)
Steel moment frames	S1	A Frame of steel columns and beams
Steel braced frames	S2	Vertical components of the lateral force-resisting system are braced frames rather than moment frames
Steel frames with unreinforced masonry infill walls	S3	The infill walls usually are offset from the exterior frame members, wrap around them, and present a smooth masonry exterior with no indication of the frame
Steel and RC composite systems	S5	Moment resisting frame of composite steel and concrete columns and beams. Usually the structure is concealed on the outside by exterior nonstructural walls
Wood structures	W	Repetitive framing by wood rafters or joists on wood stud walls. Loads are light and spans are small

dispersion) as a percentage for each of the assets or infrastructure components included in the analysis, one can make estimates of various parameters useful for the proposed analysis as the result of obtaining the Loss Exceedance Curve (LEC).

Loss Exceedance Curve, LEC, represents the annual frequency of exceedance of a specific loss. This is the most important catastrophe risk metric for risk managers, since it estimates the amount of funds required to meet risk management objectives. The LEC can be calculated for the largest event in one year or for all (cumulative) events in one year. For risk management purposes, the latter estimate is preferred, since it includes the possibility of one or more severe events resulting from earthquakes. Once calculated the Loss Exceedance Curve, it is possible to obtain other appropriate metrics for the financial analysis of the losses such as the Average Annual Loss or Pure Risk Premium for each building and for sets of buildings. The Probable Maximum Loss is obtained for the whole portfolio, that is, the entire city (Ordaz et al., 2003; Cardona et al., 2008).

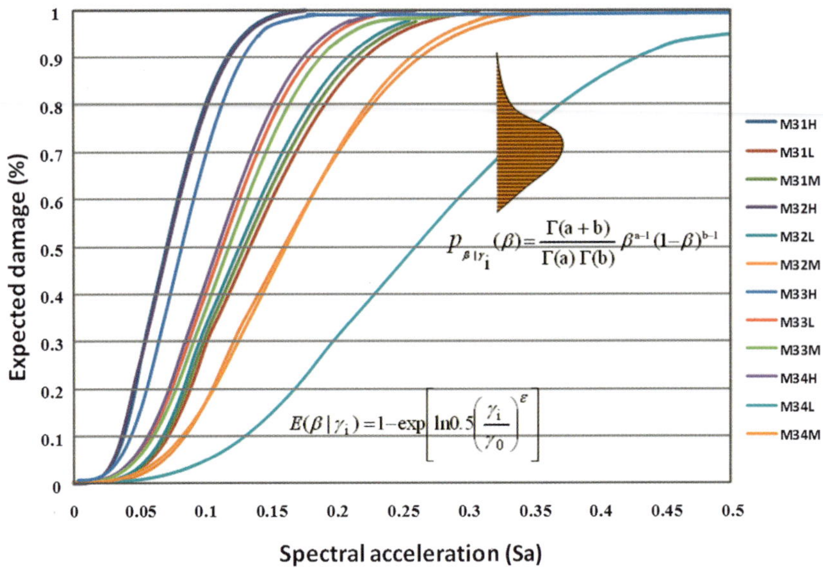

FIGURE 2.5 Vulnerability functions for unreinforced masonry buildings.

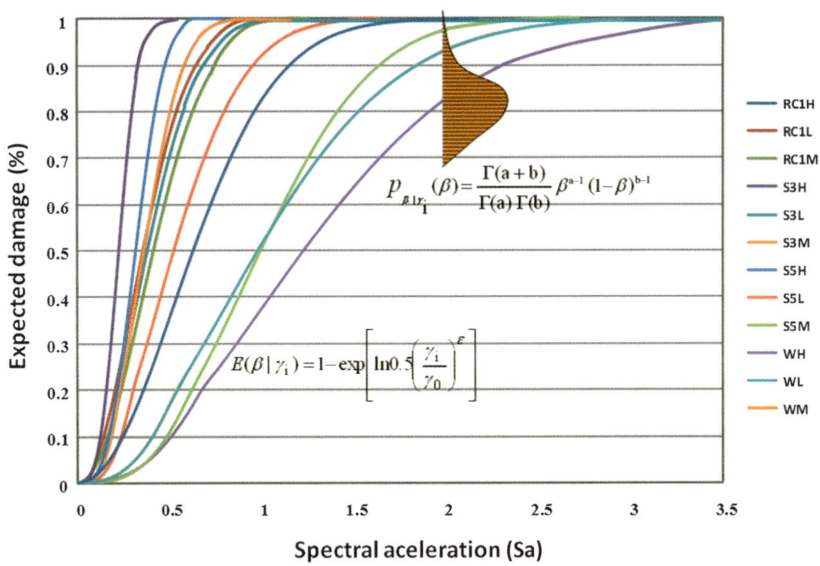

FIGURE 2.6 Vulnerability functions for reinforced concrete, steel, and wood buildings.

Average Annual Loss, AAL, is the expected loss per year. Computationally, AAL is the sum of the product of the expected losses in a specific event and the annual occurrence probability of that event, for all stochastic events considered in the loss model. The expected annual loss considers the losses of each building for all the events that can occur; supposing that the process of occurrence of hazard events is stationary and that damaged structures have their resistance immediately restored after an event (Ordaz et al., 1998; Ordaz and Reyes, 1999).

The annual occurrence frequency of events depends on the results of hazard assessments. The loss expected value, given the occurrence of a particular event, depends on the vulnerability of the exposed element.

Probable Maximum Loss, PML, represents the loss amount for a given annual exceedance frequency, or its inverse, the return period. Generally, the PML, in economic value or in percentage with regard to the return period, specifies the PML curve. The PML of an exposed base is an appraiser of the size of maximum losses reasonably expected in a set of elements exposed during the occurrence of a hazard event. Typically, PML is a fundamental data to determine the size of reserves that, for example, insurance companies or a government should maintain to avoid excessive losses that might surpass their adjustment capacity.

The Average Annual Loss for physical assets, fatalities and injuries are calculated for each building of the city. The probabilistic results for the city of Barcelona are shown in Tables 2.2–2.4; the expected annual loss is given in economic value per 1000, the number of persons is per 100,000 population and the Probable Maximum Loss (PML) values for different return periods. Figure 2.7 shows the PML curve obtained for Barcelona. Figure 2.8 shows the expected annual loss for each AEB of Barcelona. As it was previously mentioned, the expected annual loss was originally calculated building by building and Figure 2.9 shows the obtained results at this resolution. From Figure 2.8 can be seen how the expected annual loss values are highest in the old part of the city. More information about the seismic risk for the city of Barcelona can be found in Barbat et al. (2010) and additional information about a probabilistic seismic risk evaluation can be seen in Marulanda et al. (2013).

2.5 HOLISTIC RISK EVALUATION

A holistic evaluation of the disaster risk has been performed by adapting the methodology proposed by Carreño (2006) and Carreño et al. (2007a) and into the frame of the conceptual framework of the MOVE project (Birkmann et al., 2013). In this methodology, risk requires a multidisciplinary evaluation that takes into account not only the expected physical damage, the number and type of casualties, or economic losses (first order impact), but also the conditions related to social fragility and lack of resilience conditions, which favor the second order effects (indirect impact) when a seismic hazard event strikes an urban center (Carreño et al., 2007a).

In the holistic evaluation of risk using indices, risk results are achieved aggravating the physical risk by means of the contextual conditions, such as the socioeconomic fragility and the lack of resilience. Input data about these conditions at urban level are necessary to apply the method. This approach contributes to the effectiveness of risk management, inviting to the action through the identification of weaknesses of the urban center (Carreño et al., 2007a).

TABLE 2.2 Obtained Results for Physical Exposure

	Physical Exposure	
Exposed value	€ ×10^6	31,522.80
Average annual loss	€ ×10^6	72.14
	‰	2.29‰
PML		
Return period	Loss	
Years	€ ×10^6	%
50	729.35	2.31%
100	1770.16	5.62%
250	3699.35	11.74%
500	5172.26	16.41%
1000	6510.67	20.65%
1500	7021.14	22.27%

TABLE 2.3 Obtained Results for Dead People

	Dead People	
Exposed value	Inhab	1,639,880.00
Average annual loss	Inhab	28.27
	‰	0.017‰
PML		
Return period	Loss	
Years	Inhab	%
50	101.41	0.01%
100	654.30	0.04%
250	2069.97	0.13%
500	3380.29	0.21%
1000	4898.39	0.30%
1500	5799.44	0.35%

TABLE 2.4 Obtained Results for Injured People

	Injured People	
Exposed value	Inhab	1,639,880.00
Average annual loss	Inhab	113.55
	‰	0.07‰
PML		
Return period	Loss	
Years	Inhab	%
50	756.92	0.05%
100	3420.18	0.21%
250	9028.50	0.55%
500	12,590.98	0.77%
1000	15,803.45	0.96%
1500	16,903.45	1.03%

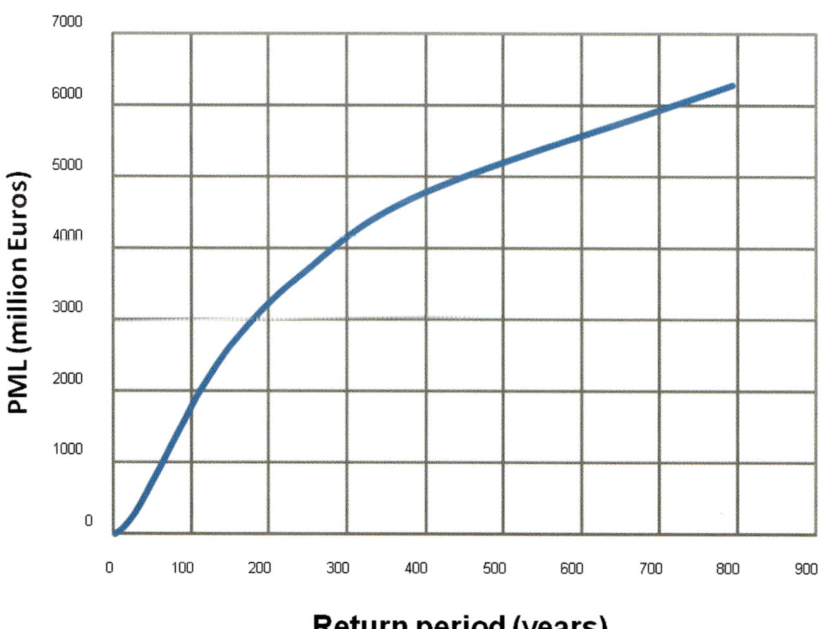

FIGURE 2.7 PML curve for Barcelona.

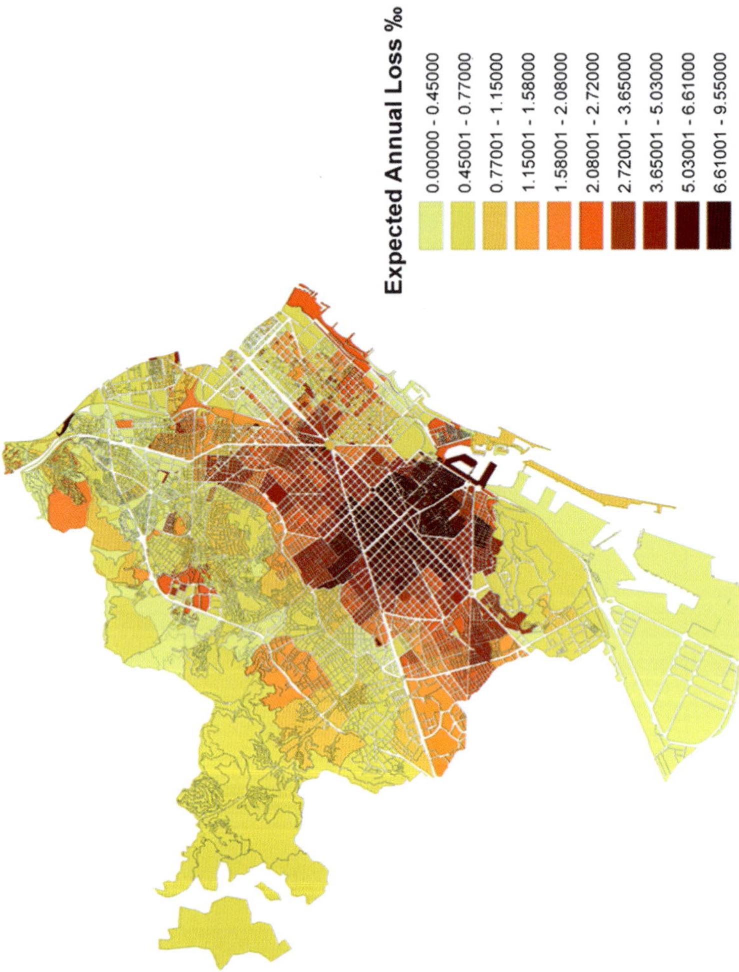

FIGURE 2.8 Expected annual loss for the AEBs of Barcelona.

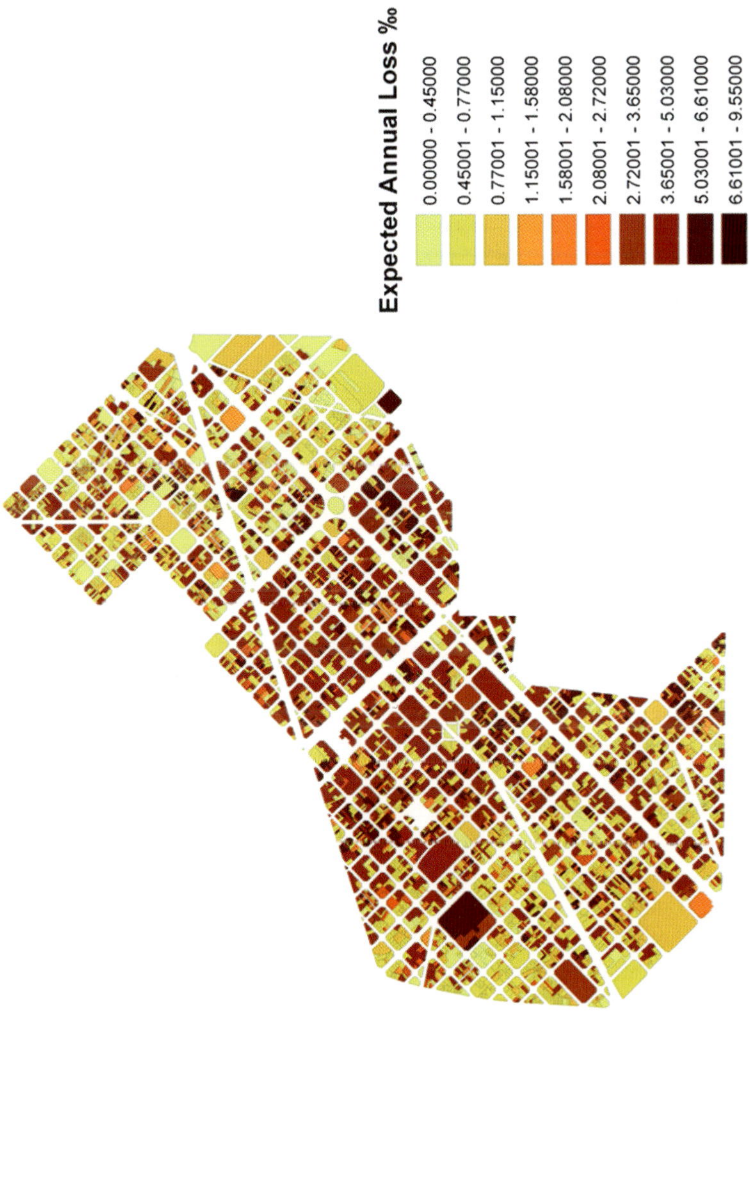

FIGURE 2.9 Expected annual loss for each building in the Eixample District of Barcelona.

The socioeconomic fragility and the lack of resilience are described by a set of indicators (related to indirect or intangible effects) that aggravate the physical risk (potential direct effects). Thus, the total risk depends on the direct effect, or physical risk, and the indirect effects expressed as a factor of the direct effects. Therefore, the total risk is expressed as follows:

$$R_T = R_F\,(1+F) \tag{2.1}$$

equation known as the Moncho's Equation in the field of disaster risk indicators, where R_T is the total risk index, R_F is the physical risk index, and F is the aggravating coefficient (Carreño et al., 2007a). This coefficient depends on the weighted sum of a set of aggravating factors related to the socioeconomic fragility, F_{FSi}, and the lack of resilience of the exposed context, F_{FRj}

$$F = \sum_{i=1}^{m} w_{FSi} F_{FSi} + \sum_{j=1}^{m} w_{FRj} F_{FRj} \tag{2.2}$$

where w_{FSi} and w_{FRj} are the weights or influences of each i and j factors and m and n are the total number of descriptors for social fragility and lack of resilience, respectively. The aggravating factors F_{FSi} and F_{FRj} are calculated using transformation functions, which are discussed in the following.

The descriptors used in this evaluation have different nature and units, the transformation functions standardize the gross values of the descriptors, transforming them into commensurable factors. Figure 2.10 shows a model for the transformation functions used by the methodology in order to calculate the risk and aggravating factors. They are membership functions for high level of risk and high aggravating level for each. In Figure 2.10, the x-axis values are of the descriptors while the value of the factor (physical risk or aggravation) is in the y-axis, taking values between 0 and 1, were 0 is the non-membership and 1 is the total membership. The limit values, $X_{\min}$ and $X_{\max}$, are defined taking into account the expert opinions and information about past disasters. In the case of the descriptors of lack of resilience, the function has the inverse shape; the higher value of the indicator gives lower value of aggravation. The weights w_{FSi} and w_{FRj} represent the relative importance of each factor and are calculated by means of the Analytic Hierarchy Process (AHP) on the basis of local expert opinions. The AHP is used to derive ratio scales from both discrete and continuous paired comparisons (Saaty, 1991; Carreño et al., 2007; Carreño, 2006). These comparisons are performed by local experts and the researcher team.

The physical risk, R_F, is evaluated in the same way, by using the following equation:

$$R_F = \sum_{i=1}^{p} w_{RFi} F_{RFi} \tag{2.3}$$

Figure 2.11 shows the process of calculation of the total risk index for the units of analysis, which could be districts, municipalities, communes, or

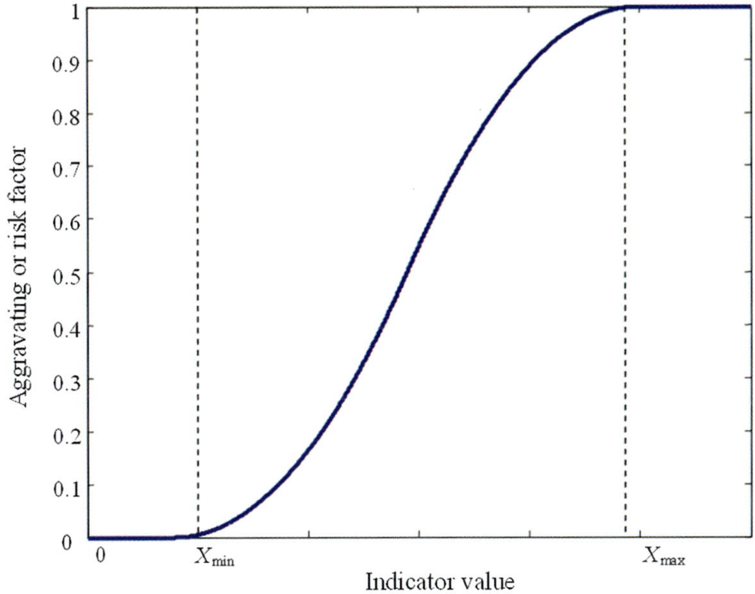

FIGURE 2.10 Model of the transformation functions.

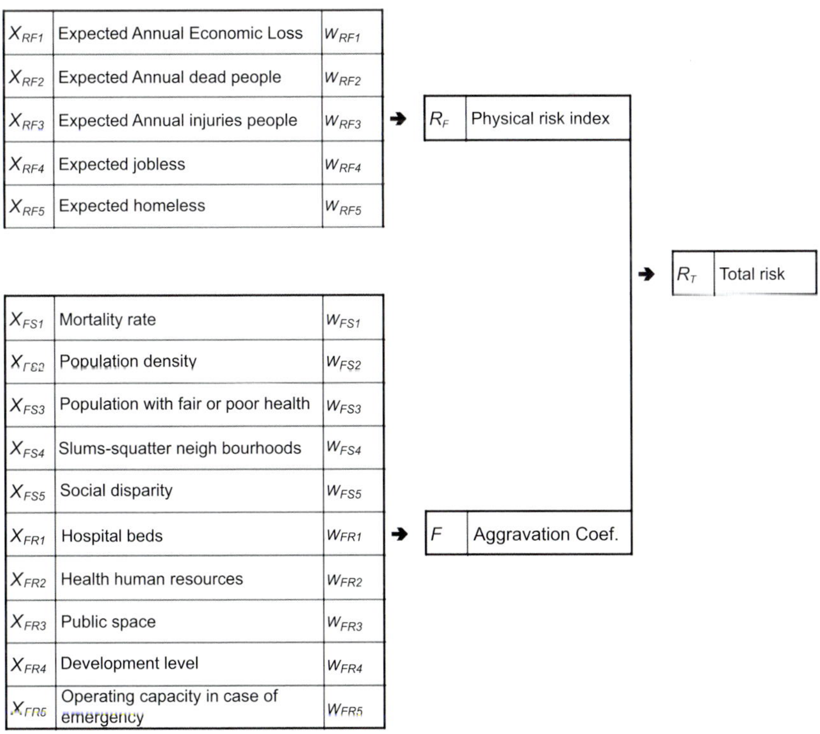

FIGURE 2.11 Descriptors of the physical risk, social fragility, and lack of resilience and their weights.

localities, starting from the descriptors of physical risk, X_{RFi}, and the descriptors of the aggravating coefficient F, that is, X_{FSi} and X_{RFi}, using the weights w_{RFi}, w_{FSi}, and w_{FRi} of each descriptor.

This case study is a focus in the holistic evaluation of the seismic risk. Figure 2.11 shows also the descriptors used to describe the physical risk, the social fragility, and the lack of resilience for the case study of Barcelona. These descriptors were chosen as the most significant for each category, and according to available information of the city. The descriptors of physical risk correspond to the obtained results of the probabilistic evaluation of seismic risk of the previous section of this chapter. The descriptors of social fragility and lack of resilience correspond to available information of the city.

Also, the robustness of this methodology has been studied assessing the uncertainty of values and sensitivity to change of values, weights, and transformation functions (Marulanda et al., 2009; Carreño et al., 2009). The methodology is not excessively sensitive to slight variations in the input data and to small changes in the modeling parameters, such as weights and transformation functions. The results do not show important or extreme changes. If the range of variation of data and parameters is reasonable, as it is in the case of seismic risk, the results of the model will be stable and reliable. Detailed information about this evaluation method can be founded in references (Carreño et al., 2007a; Carreño, 2006; Barbat et al., 2011).

The holistic evaluation of risk has been done by following the exposed methodology. Figure 2.12 shows the obtained results of the physical risk index, R_F, for the AEB's of Barcelona; Figure 2.13 shows the ranking of the average values for the districts of the city. These results give the highest values of physical risk in areas of the districts of Ciutat Vella and Eixample; these areas correspond to the older areas in the city. The smaller values are in areas of the Nou Barris District and Horta-Guinardo districts.

The results of the aggravating coefficient are shown in Figure 2.14 for each district of the city; Figure 2.15 shows the ranking of these results. The district of Ciutat Vella has the worst aggravating situation according to the characteristics of social fragility and lack of resilience, the best situation is for the Sarria-Sant Gervasi and Les Corts districts.

Figure 2.16 shows the results of the total risk index, R_T, for the AEB's of Barcelona. Figure 2.17 shows a detail of the results for the AEB's in the Eixample district; and Figure 2.18 shows the ranking of the average values of R_T for the district of Barcelona.

2.6 EVALUATION OF THE RISK MANAGEMENT PERFORMANCE

For management purposes, the risk assessment should be to improve the decision-making process in order to contribute to the effectiveness of risk management, calling for action and identifying the weaknesses of the exposed elements and their evolution over time (Carreño et al., 2007b).

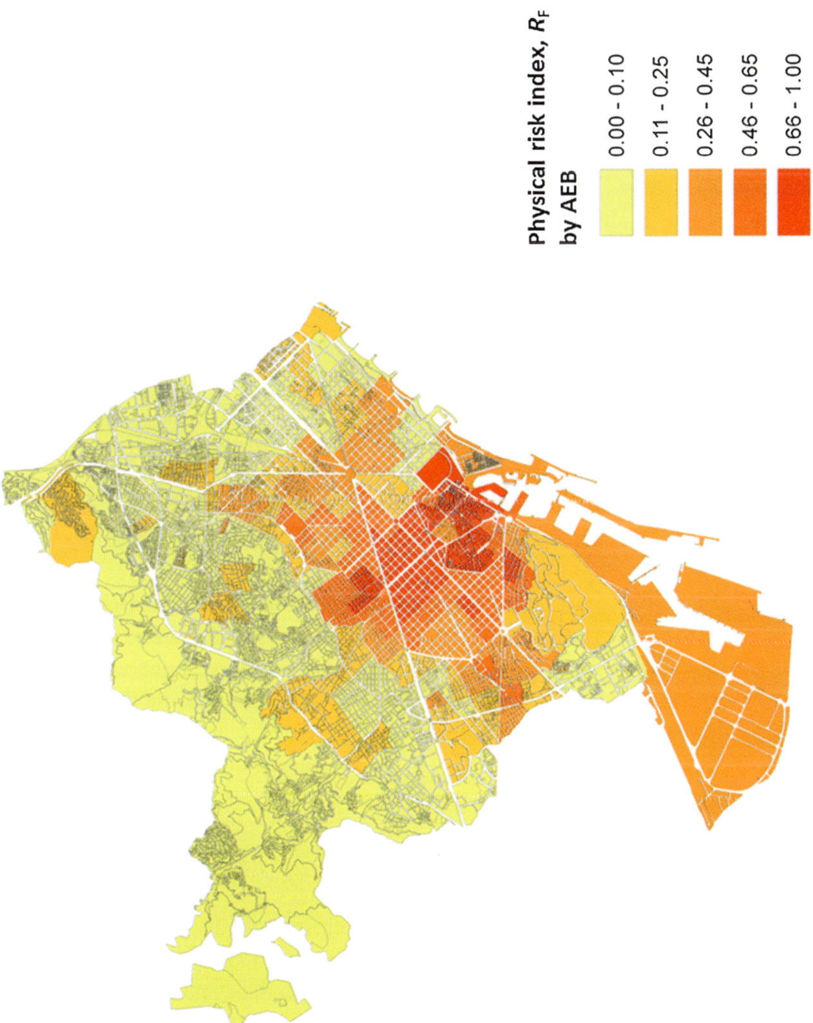

FIGURE 2.12 Physical risk index.

The evaluation of the risk management performance has been performed by using the methodology proposed by IDEA (2005), Carreño (2006) and Carreño et al. (2007b). This methodology calculates the Risk Management Index, RMI, which brings together a group of indicators that measure risk management performance and effectiveness. These indicators reflect the organizational development capacity and institutional actions taken to reduce vulnerability and losses in a given area, to prepare for crisis, and to recover efficiently from disasters; they are evaluated on the basis of the local expert opinions. This index is designed to assess risk management performance. It provides a quantitative measure of management based on predefined qualitative targets or benchmarks that risk management efforts should aim to achieve.

The RMI was designed to assess risk management performance and, by this way, its effectiveness. It provides a quantitative measure of management based on predefined qualitative targets or benchmarks that risk management efforts should aim to achieve. The design of the RMI involved establishing a scale of achievement levels (Davis, 2003; Masure, 2003) or determining the distance between current conditions and an objective threshold or conditions in a reference country (Munda, 2003). The RMI was constructed by quantifying four public policies, each of which having six indicators. Risk identification index, RMI_{RI}, is a measure of individual perceptions, of how those perceptions are understood by society as a whole, and the objective assessment of risk. Risk reduction index, RMI_{RR}, involves prevention and mitigation measures. Disaster management index, RMI_{DM}, involves measures of response and recovery, and governance and financial protection, RMI_{FP}, measures the degree of institutionalization and risk transfer. The four public policies and their indicators were defined after an agreement with several stakeholders and evaluators. Any country or city could redefine them according to own specificities, whereas the parameters are maintained the same in the distinct evaluations over time, in order to make a consistent follow-up of the risk management. The RMI is defined as the average of the four composite indicators

$$RMI = (RMI_{RI} + RMI_{RR} + RMI_{DM} + RMI_{FP}) / 4 \quad (2.4)$$

Six indicators are proposed for each public policy. Together, these serve to characterize the risk management performance of a country, region, or city. Using a larger number of indicators could be redundant and unnecessary and could make the weighting of each indicator difficult. Following the performance evaluation of risk management method proposed by Carreño et al. (2007a), the valuation of each indicator is based on five performance levels (low, incipient, significant, outstanding, and optimal) that correspond to a range from 1 (low) to 5 (optimal). Examples of these performance levels can be seen as follows for the

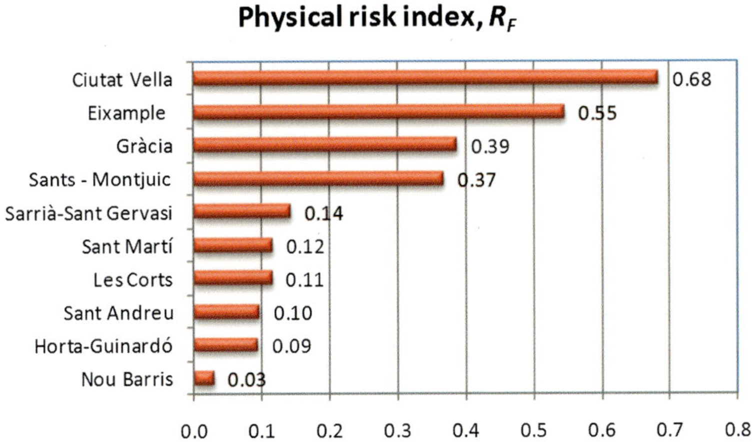

FIGURE 2.13 Ranking of the average physical risk index for the districts of Barcelona.

case of "Public information and community participation" and "Risk consideration in land use and urban planning".

2.6.1 RI5: Public Information and Community Participation

1. Sporadic information on risk management in normal conditions and more frequently when disasters occur.
2. Press, radio, and television coverage oriented toward preparedness in case of emergency: production of illustrative materials on dangerous phenomena.
3. Frequent opinion programs on risk management issues at the national and local levels: guidelines for vulnerability reduction; and work with communities and NGOs.
4. Generalized diffusion and progressive consciousness; conformation of some social networks for civil protection and NGOs that explicitly promote local risk management issues and practice.
5. Wide scale participation and support from the private sector for diffusion activities. Consolidation of social networks and notable participation of professionals and NGOs at all levels.

2.6.2 RR1: Risk Consideration in Land Use and Urban Planning

1. Consideration of some means for identifying risk, and environmental protection in physical planning.
2. Promulgation of national legislation and some local regulations that consider some hazards as a factor in territorial organization and development planning.

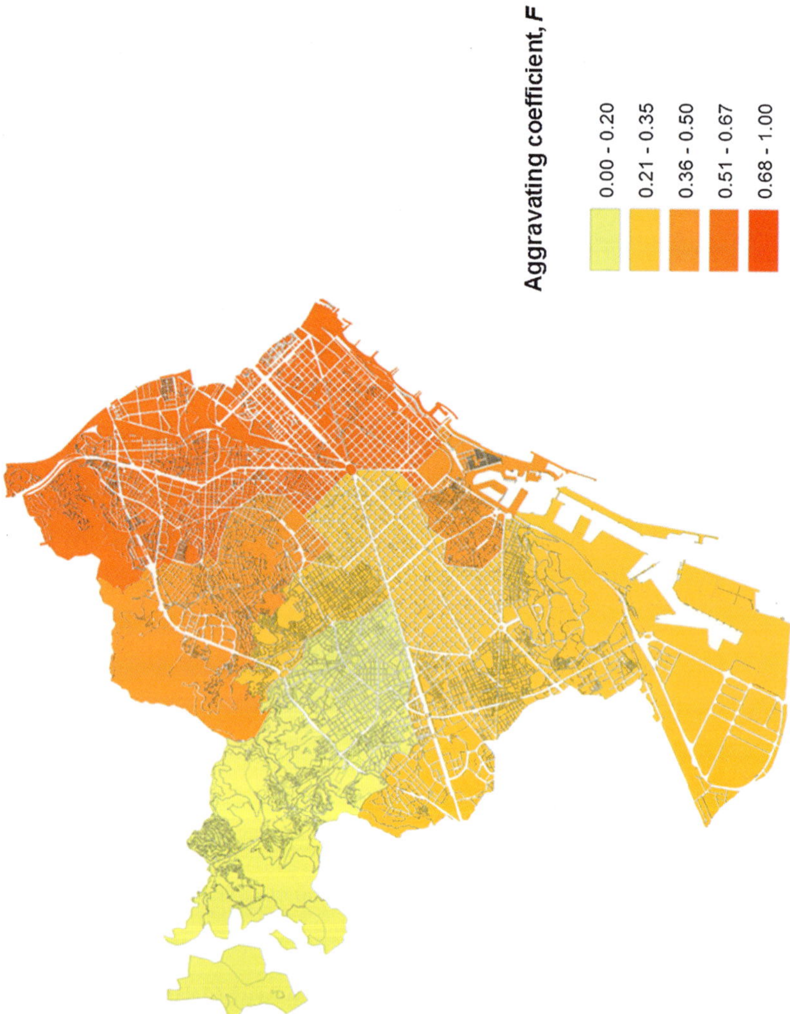

FIGURE 2.14 Aggravating coefficient for the Barcelona districts.

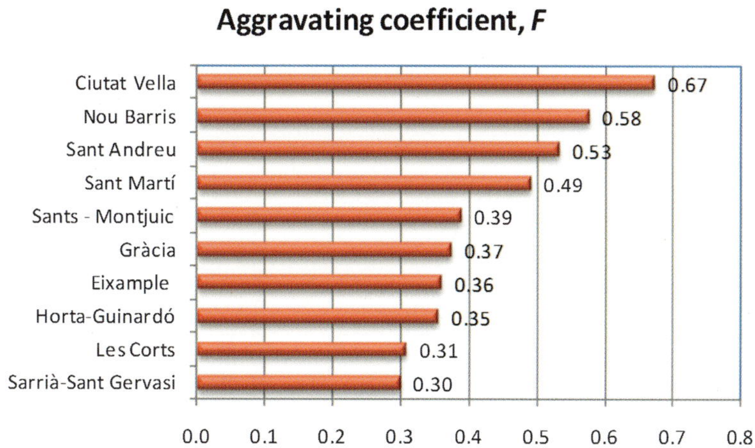

FIGURE 2.15 Ranking of the aggravating coefficient of the districts of Barcelona.

3. Progressive formulation of land use regulations in various cities that take into account hazards and risks; obligatory de-sign and construction norms based on microzonations.
4. Wide ranging formulation and updating of territorial organization plans with a preventive approach in the majority of municipalities. Use of microzonations with security ends. Risk management incorporation into sectorial plans.
5. Approval and control of implementation of territorial organization and development plans that include risk as a major factor and the respective urban security regulations.

Computationally, these performance levels are represented by using the fuzzy sets theory. This methodological approach permits the use of each reference level simultaneously as a performance target and allows for comparison and identification of results or achievements. Government efforts at formulating, implementing, and evaluating policies should bear these performance targets in mind. Figure 2.19 shows the indicator components of each one of the public policies evaluated by the RMI.

In the case of Barcelona, the indicators have been evaluated by local experts from different disciplines; the obtained results presented as follows are calculated taking the average of the evaluations done. Table 2.5 shows the qualifications for the risk identification indicators and their correspondent weights.

As an example, the development level of the indicator RI2 *(Hazard monitoring and forecasting)* in 2010 was qualified as 4 (*outstanding*). The *Servei Meteorològic de Catalunya* (http://www.meteo.cat/servmet/smr/index.html), for threats of meteorological origin, has a set of equipment for monitoring which includes a network of meteorological observers (XOM), a network of automatic

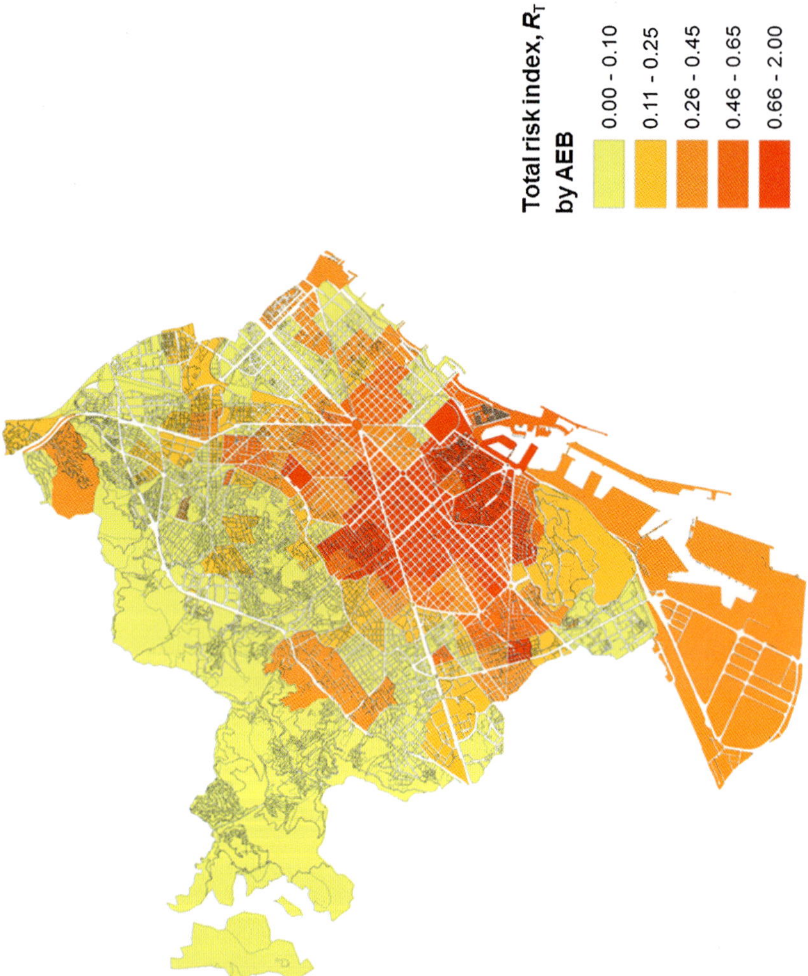

FIGURE 2.16 Total risk index for Barcelona.

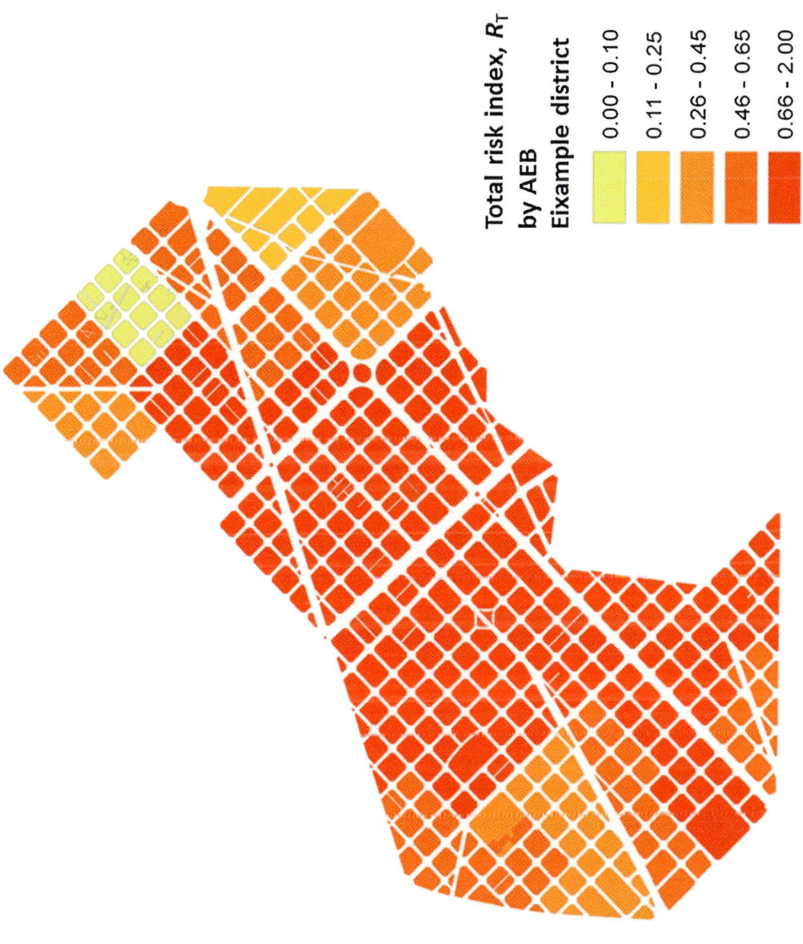

FIGURE 2.17 Detail of the total risk for the Eixample district.

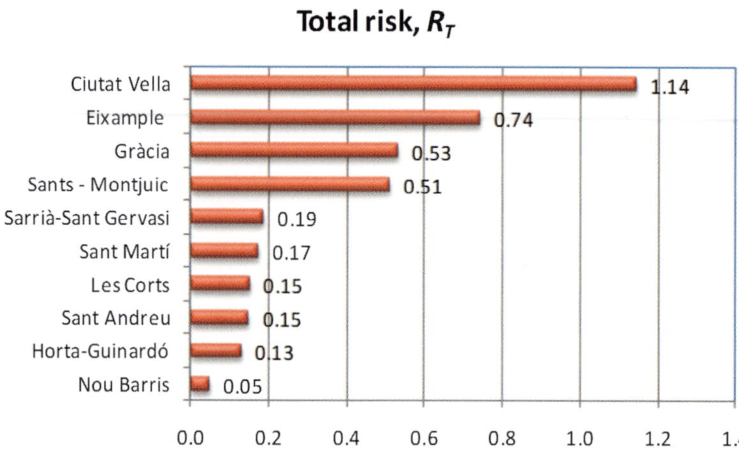

FIGURE 2.18 Ranking of the average total risk index for the districts of Barcelona.

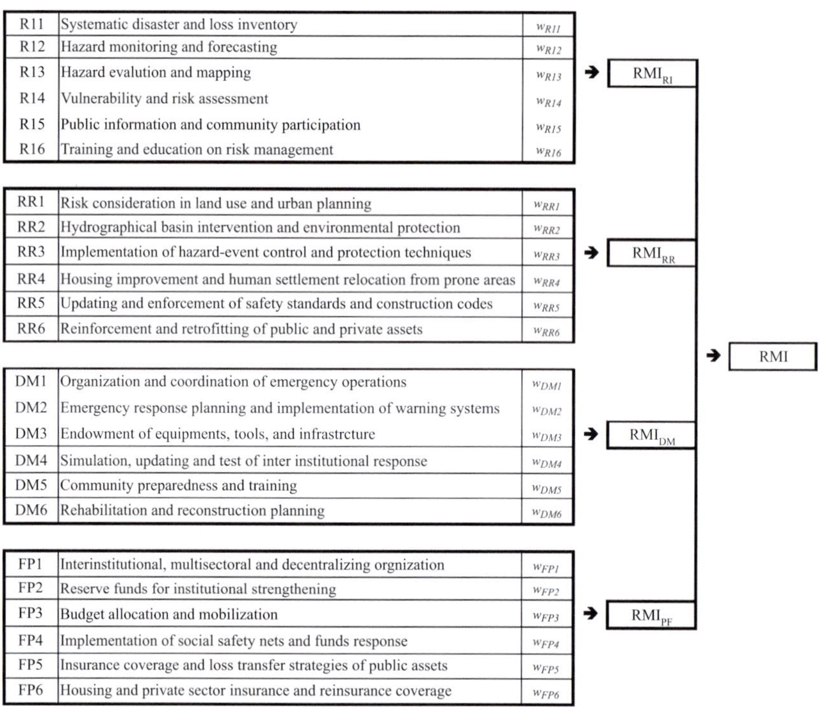

FIGURE 2.19 RMI and its component indicators (IDEA, 2005).

TABLE 2.5 Qualifications for the Indicators of Risk Identification (RI)

	RI1	RI2	RI3	RI4	RI5	RI6
1990	2	2	2	1	1	1
1995	3	3	2	2	1	1
2000	3	4	3	3	1	1
2005	3	4	3	3	2	1
2010	3	4	4	4	2	1
Weight	22	18	18	20	12	10

weather stations (XEMA), a radar network (XRAD), a network of radio exploration and Meteosat satellite receivers. Some of the functions of this institution are:

1. To schedule, to implement and to manage the system of meteorological phenomena monitoring and forecasting, and to make the exploitation and dissemination in Catalonia
2. To forecast, to watch and to monitor the meteorological risk situations, in coordination with the Emergency Center of Catalonia (CECAT) with a view to improving the effectiveness of activities and ensure appropriate communication to users in may be affected by the system of alerts that is determined by regulatory proceedings
3. To permanently provide, when meteorological risk situations have been forecasted, official meteorological advice to the administrations and institutions responsible for civil protection in Catalonia and to collaborate in such cases, if any, with the meteorological authority of the State
4. To promote research activities in the field of meteorology and climatology and encourage the development of products and services in this area
5. To organize training activities and dissemination in the field of meteorology
6. To carry out studies to improve the understanding of climate and weather in Catalonia
7. To advise and to assist the government agencies on matters related to the study of climate and climate change, in coordination with relevant agencies in this area
8. To advise and to assist the competent public authorities, and to collaborate with them in monitoring and prediction of phenomena and air pollution episodes
9. To analyze and to monitor the characteristics of column ozone in Catalonia in relation to the evolution of stratospheric ozone, and inform the population
10. To study and to analyze climate change in Catalonia and to participate and to collaborate in research on this matter to carry out various groups in Catalonia

With regard to the hazards of seismic and geological origin, the Institut Geològic de Catalunya, IGC (www.igc.cat), maintains a regional seismic network with the aim to monitor the seismicity in Catalonia and nearby areas (Eastern Pyrenees and the Mediterranean Sea). The network is composed of analog stations, as well as digital stations. IGC has a network of accelerometers, most of them with continues transmission in real time.

Tables 2.6–2.8 show the qualification for the indicators of risk reduction, disaster management and financial protection and their correspondent weights. Table 2.9 shows the obtained results for Barcelona.

Figures 2.20 and 2.21 show the obtained results for the RMI and its components for the studied years.

2.7 CONCLUSIONS

This evaluation study perfectly fits with the conceptual framework presented in the Chapter 2 of this book. This evaluation includes several elements involved in the conceptual framework such as: seismic hazard, exposure, susceptibility and fragility for physical, social, economical and institutional issues, lack of resilience, risk, risk governance and risk management.

This study develops a risk evaluation with a multidisciplinary approach that takes into account not only the expected physical damage, the number and type of casualties or economic losses, but also other social, organizational, and institutional issues related to the development of communities that contribute to the creation of risk. Therefore, the potential negative consequences are not only related to the effects of the hazardous event as such, but also to the capacity to absorb the effects and the control of its implications in a given geographical area.

This holistic evaluation of risk facilitates the integrated risk management by the different stakeholders involved in risk reduction decision-making. It permits the follow-up of the risk situation and the effectiveness of the prevention and mitigation measures can be easily achieved. Results can be verified and the mitigation priorities can be established as regards the prevention and planning actions to modify those conditions having a greater influence on risk in the city.

TABLE 2.6 Qualifications for the Indicators of Risk Reduction (RR)

	RR1	RR2	RR3	RR4	RR5	RR6
1990	1	1	2	1	2	1
1995	2	2	2	2	3	1
2000	2	3	3	3	3	1
2005	3	3	4	3	3	1
2010	3	3	4	3	4	2
Weight	16	14	22	15	17	16

Once the results have been expressed in graphs for each locality or district, it is easy to identify the most relevant aspects of the total risk index, with no need for further analysis and interpretation of results.

The holistic evaluation was done with the basis of a probabilistic analysis of the physical seismic risk. Probabilistic techniques are used to calculate the

TABLE 2.7 Qualifications for the Indicators of Disaster Management (DM)

	DM1	DM2	DM3	DM4	DM5	DM6
1990	2	1	1	1	1	1
1995	2	1	1	1	1	1
2000	3	2	2	1	1	1
2005	3	3	2	1	1	1
2010	3	3	3	2	1	2
Weight	27	20	19	11	11	12

TABLE 2.8 Qualifications for the Indicators of Financial Protection (FP)

	FP1	FP2	FP3	FP4	FP5	FP6
1990	1	1	1	2	1	1
1995	1	1	1	2	1	1
2000	1	1	1	2	2	2
2005	1	2	1	3	2	2
2010	2	3	2	3	2	2
Weight	20.5	15.4	17.2	12.8	17.9	16.2

TABLE 2.9 Obtained Results for the RMI and Its Components

	RMI_{RI}	RMI_{RR}	RMI_{DM}	RMI_{FP}	RMI
1990	12.61	13.29	13.23	10.77	12.47
1995	34.2	30.83	13.23	10.77	22.26
2000	49.97	34.03	35.63	11.87	32.88
2005	47.21	50.9	35.84	28.07	40.5
2010	48.07	51.24	37.65	34.83	42.95

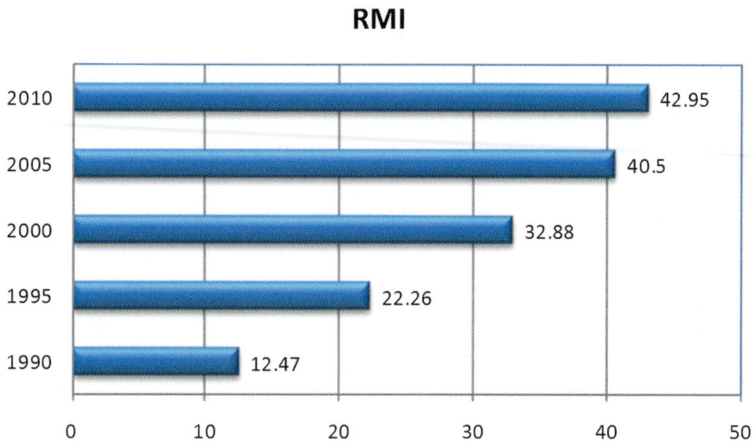

FIGURE 2.20 Results of the RMI for 1990, 1995, 2000, 2005, and 2010.

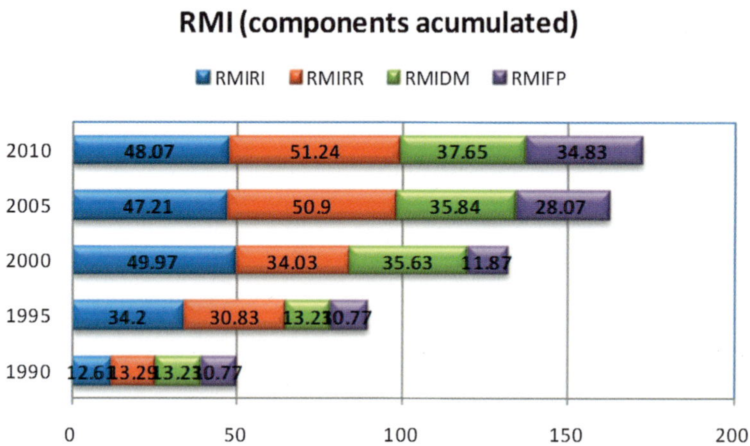

FIGURE 2.21 Results of the components RMI_{RI}, RMI_{RR}, RMI_{DM}, and RMI_{FP}.

Probable Maximum Loss (PML) for different return periods and the Average Annual Loss (AAL) also known as technical risk premium. The CAPRA platform was used to calculate the mentioned metrics for the city of Barcelona, Spain, to estimate scenarios of damages and losses.

The evaluation of the disaster risk management index, RMI, permits a systematic and quantitative benchmarking of the city during different periods. This index not only enables the depiction of disaster risk management at urban level, allowing the creation of risk management performance benchmarks in order to establish performance targets for improving management effectiveness. The obtained results for Barcelona shows an evolution of the disaster risk management in the last 20 years, but also reveals the pending work to be done in this area.

REFERENCES

Ambraseys, N.N., Simpson, K.A., Bommer, J.J., 1996. Prediction of horizontal response espectra in Europe. Earthquake Engineering and Structural Dynamics 25, 375–400.

Barbat, A.H., Carreño, M.L., Cardona, O.D., Marulanda, M.C., 2011. Evaluación holística del riesgo sísmico en zonas urbanas. Revista Internacional de Métodos Numéricos para Cálculo y Diseño en Ingeniería 27, 3–27.

Barbat, A.H., Carreño, M.L., Pujades, L.G., Lantada, N., Cardona, O.D., Marulanda, M.C., 2010. Seismic vulnerability and risk evaluation methods for urban areas. A review with application to a pilot area. Structure and Infrastructure Engineering 6 (1–2), 17–38.

Birkmann, J., Cardona, O.D., Carreño, M.L., Barbat, A.H., Pelling, M., Schneiderbauer, S., Kienberger, S., Keiler, M., Alexander, D., Zeil, P., Welle, T., 2013. Framing vulnerability, risk and societal responses: the MOVE framework. Natural Hazards 67 (2), 193–211.

Cardona, O.D., Ordaz, M.G., Marulanda, M.C., Barbat, A.H., 2008. Estimation of probabilistic seismic losses and the public economic resilience—an approach for a macroeconomic impact evaluation. Journal of Earthquake Engineering 12 (sup2), 60–70.

Carreño, M.L., 2006. Técnicas innovadoras para la evaluación del riesgo sísmico y su gestión en centros urbanos: Acciones ex ante y ex post (Thesis doctoral). Universidad Politécnica de Cataluña, Barcelona.

Carreño, M.L., Cardona, O.D., Barbat, A.H., 2007a. Urban seismic risk evaluation: a holistic approach. Natural Hazards 40 (1), 137–172.

Carreño, M.L., Cardona, O.D., Barbat, A.H., 2007b. Disaster risk management performance index. Natural Hazards 41 (1), 1–20.

Carreño, M.L., Cardona, O.D., Marulanda, M.C., Barbat, A.H., 2009. Holistic urban seismic risk evaluation of megacities: application and robustness. Geotechnical, Geological and Earthquake Engineering, Springer 7, 167–183.

Cid, J., Susagna, T., Goula, X., Chavarria, L., Figueras, S., Fleta, J., Casas, A., Roca, A., 2001. Seismic zonation of barcelona based on numerical simulation of site effects. Pure Applied Geophysics 158, 2559–2577.

Davis, I., 2003. The Effectiveness of Current Tools for the Identification, Measurement, Analysis and Synthesis of Vulnerability and Disaster Risk. National University of Colombia, Manizales. IDB/IDEA Program of Indicators for Risk Management. http://idea.unalmzl.edu.co.

Egozcue, J.J., Barbat, A.H., Canas, J.A., Miquel, J., Banda, E., 1991. A method to estimate occurrence probabilities in low seismic activity regions. Earthquake Engineering and Structural Dynamics 20 (1), 43–60.

Geoter, 2008. Calcul probabilista de la perillositat sísmica a Catalunya en acceleracions espectrals. Institut Geològic de Catalunya, Barcelona, España. Document GTB/IGC/0508-001.

ICC/CIMNE, 2004. An advanced approach to earthquake risk scenarios with applications to different European towns. WP08 Application to Barcelona. RISK-UE Project.

IDEA, 2005. System of Indicators for Disaster Risk Management: Main Technical Report. Universidad Nacional de Colombia, Manizales. IDB/IDEA Programme of Indicators for Disaster Risk Management.

Irizarry, J., Lantada, N., Pujades, L.G., Barbat, A.H., Goula, X., Susagna, T., Roca, A., 2011. Ground-shaking scenarios and urban risk evaluation of Barcelona using the Risk-UE capacity spectrum based method. Bulletin of Earthquake Engineering 9 (2), 441–466.

Lantada, N., 2007. Evaluación del riesgo sísmico mediante métodos avanzados y técnicas gis. Aplicación a la ciudad de Barcelona (Doctoral dissertation). Universidad Politécnica de Cataluña, Barcelona, Spain.

Lantada, N., Pujades, L.G., Barbat, A.H., 2009. Vulnerability index and capacity spectrum based methods for urban seismic risk evaluation. A comparison. Natural Hazards 51 (3), 501–524.

Lantada, N., Irrizari, J., Barbat, A.H., Goula, X., Roca, A., Susagna, T., Pujades, L.G., 2010. Seismic hazard and risk scenarios for Barcelona, Spain, using the risk-UE vulnerability index method. Bulletin of Earthquake Engineering 8 (1), 01–229.

Marulanda, M.C., Cardona, O.D., Barbat, A.H., 2009. Robustness of the holistic seismic risk evaluation in urban centers using the USRi. Natural Hazards 49, 501–516.

Marulanda, M.C., Carreño, M.L., Cardona, O.D., Ordaz, M.G., Barbat, A.H., 2013. Probabilistic earthquake risk assessment using CAPRA: application to the city of Barcelona, Spain. Natural Hazards 69 (1), 59–84.

Masure, P., 2003. Variables and Indicators of Vulnerability and Disaster Risk for Land-use and Urban or Territorial Planning, IDB/IDEA Program of Indicators for Disaster Risk Management. National University of Colombia, Manizales. http://idea.unalmzl.edu.co.

Munda, G., 2003. Methodological Exploration for the Formulation of a Socio-Economic Indicators Model to Evaluate Disaster Risk Management at the National and Sub-National Levels. A Social Multi-criterion Model. IDB/IDEA Program of Indicators for Disaster Risk Management. National University of Colombia, Manizales. http://idea.unalmzl.edu.co.

Ordaz, M.G., Miranda, E., Reinoso, E., Pérez-Rocha, L.E., 1998. Seismic Loss Estimation Model for México City. Universidad Nacional Autónoma de México. México DF., México.

Ordaz, M.G., Reyes, C., 1999. Seismic hazard in Mexico City: observations vs. computations. Bulletin of the Seismological Society of America 89 (5), 1379–1383.

Ordaz, M.G., 2000. Metodología para la Evaluación del Riesgo Sísmico Enfocada a la Gerencia de Seguros por Terremoto. Universidad Nacional Autónoma de México, México D.F.

Ordaz, M.G., Santa-Cruz, S., 2003. Computation of Physical Damage to Property Due to Natural Hazard Events. IDB/IDEA Program of Indicators for Risk Management. National University of Colombia, Manizales.

Ordaz M, Aguilar A, Arboleda J., 2007. Program for computing seismic hazard, CRISIS, Universidad Nacional Autónoma de México, Instituto de Ingeniería.

Saaty, T.L., Vargas, L.G., 1991. Prediction, Projection, and Forecasting: Applications of the Analytical Hierarchy Process in Economics, Finance, Politics, Games, and Sports. Kluwer Academic Publishers, Boston, USA.

Secanell, R., Goula, X., Susagna, T., Fleta, J., Roca, A., 2004. Seismic hazard zonation of Catalonia, Spain integrating uncertainties. Journal of Seismology 8 (1), 24–40.

Chapter 3

Spatial and Holistic Assessment of Social, Economic, and Environmental Vulnerability to Floods—Lessons from the Salzach River Basin, Austria

Stefan Kienberger, Diana Contreras and Peter Zeil
Department of Geoinformatics – Z_GIS, University of Salzburg, Salzburg, Austria

Chapter Outline

3.1 INTRODUCTION AND BACKGROUND

In 2002, 2005 and 2013, the Salzach river catchment has been severely affected by various flood events. In the context of the EC FP7 research project MOVE, the objective was to assess the vulnerability to floods in a spatially explicit way, being independent from administrative boundaries through the application of novel regionalization approaches. Furthermore, different dimensions of the commonly developed MOVE vulnerability and risk framework (Birkmann et al., 2013) were represented spatially to explore the applicability.

Assessment of Vulnerability to Natural Hazards. http://dx.doi.org/10.1016/B978-0-12-410528-7.00003-5

The case study area embraces the Austrian part of the Salzach river catchment, the Salzach being one of the main tributaries of the Inn River (as well as a subcatchment of the Danube River Basin). Along 225 km, the Salzach drains a large part of the Eastern Alps in Austria and collects waters from a catchment area of 6.649 km² within an altitude range of more than 3000 m (see Fig. 3.1). The alpine regime at the headwaters and the middle reaches and pre-alpine at its lower course characterize the catchment.

The majority of the population lives in the city of Salzburg (approx. 150,000 inhabitants) and its urban fringe. The total population of the study area comprises approximately 454,000 inhabitants. The area is highly dynamic in regard to its economic and settlement developments, putting high pressure on adequate spatial planning policies in an area with a limited amount of suitable land. Important traffic corridors follow the Salzach river in a north-south direction, and alpine areas are intensively used by tourism activities as well as for energy production.

Climate zones consist of high mountain regimes in the upstream areas, and moderate continental conditions in the lowlands. The northern areas, in which the city of Salzburg is located, receive a large amount of precipitation due to

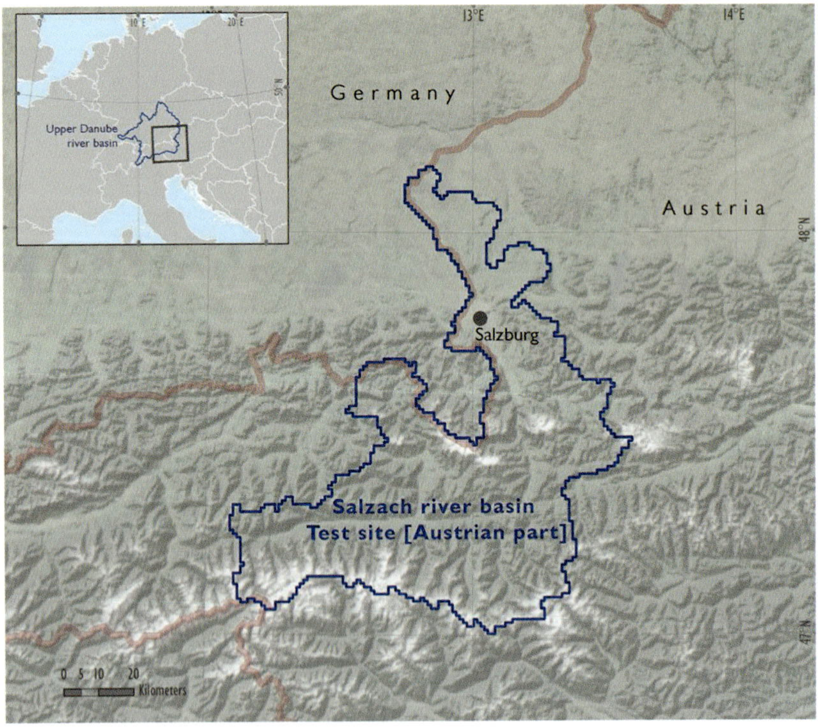

FIGURE 3.1 Extent of the Salzach river basin test case area. The study area only comprises the Austrian part of the Salzach river basin, due to improved data availabilities (such as detailed census data) in Austria.

the blocking effect of the Alps, and thus reaches up to 1400 mm in the northern-alpine Salzach valley and up to 1600 mm at the Sonnblick observatory.

To achieve the objective to protect people against floods, Austria is currently implementing flood hazard mappings, improvement of river regulations, and technical flood protection measures along with sophisticated early warning systems. Inspite of these measures, floods are not prevented per-se. They allow for enough time to prepare, warn, and evacuate people and minimize economic damage and loss of human lives based on specific flood protection activities.

This case study application aims to integrate disaggregated census data as well as other ancillary data representing the social, economic, and environmental dimensions of vulnerability. We aimed to model homogenous vulnerability regions which significantly extents the concept of spatial composite indicators or multicriteria assessment approaches. As such, homogenous regions in terms of their degree as well as their inherent characteristic of vulnerability should facilitate the improved identification of place-specific intervention measures for disaster risk reduction. This chapter summarizes key steps of the applied method as well discusses key indicators and reflects on the results achieved. For more details on methodology, we refer to the references included below.

3.1.1 Flood Hazards in the Salzach Catchment

Major hazards occurring in the case study area include floods, landslides, avalanches, debris flows flash floods, torrential rainfall, and severe storms. The vulnerability assessment in the context of the MOVE project focused only on the river flood hazard, but applied to different dimensions of vulnerability.

Since many centuries, the Salzach River has been exposed to flooding causing widespread damage not only to the city but also to rural areas. One of the worst floods occurred in 1571, with other significant events in 1789, 1899, and 1920. The most recent floods caused severe damage in 2002 (with a 2300 m^3/s flow rate in the city of Salzburg representing an HQ100), 2005, and recently in 2013 (again an HQ100 event).

Extreme precipitation and widespread thunderstorm activity with several variations in rainfall intensity were observed from northern Germany to Austria in the beginning of August 2002, exceeding monthly average values in the area. Particularly, two events from August 6th to 7th and from 12th to 13th caused severe floods in Central Europe (Ulbrich et al., 2006). Salzburg was declared a disaster zone, and even Vienna was considered under threat. The Danube was closed to all shipping, as the river swelled to a near 100-year high. According to the emergency services, 8000 workers and volunteers were available to support evacuation activities. For the 2013 events, the water gauge in Salzburg City measured 8.51 m and was thus 15 cm higher than during the catastrophic floods of 2002. The event had a major effect on transportation networks, although the total damage was much less than in 2002. This has to do with mitigation measures implemented since then.

Besides direct impacts of the flood events, also secondary effects could be observed during the 2005 event. An isolated case was studied by Schmid et al.

(2005) with respect to an outbreak of gastrointestinal illness due to a norovirus infection among a group of American tourist. They were exposed to contaminated water in a hotel, flooded on the day of their arrival in Salzburg. According to their testimony, they helped the hotel staff to clean up. Later, fire fighters who pumped the water from the affected hotel fell ill with vomiting or diarrhea. Nevertheless, although floods are usually associated with an increased risk of infection, the risk remains low, unless it is associated with population displacements or water sources being contaminated. However, none of these situations took place in Salzburg so one may conclude that the infections were contracted through direct contact with polluted water (ibid.).

Such diseases may be specific cases of secondary effects in the case study. Rainfall increases during the events and the flood itself can also trigger further secondary events such as landslides, as was severely experienced during the very recent June 2013 event. The landslides significantly affected transportation networks (roads, railways) and restricted access to the Pinzgau district, as as well as causing severe damages to houses in other areas.

3.2 ASSESSING VULNERABILITY—FROM CONCEPTS TO MEASURABLE AND MEANINGFUL SPATIAL UNITS

The assessment of vulnerability with its underlying domains was carried out at catchment scale which refers to the subnational scale as identified in the MOVE framework (Birkmann et al., 2013, Figure 3.2). The assessment targeted

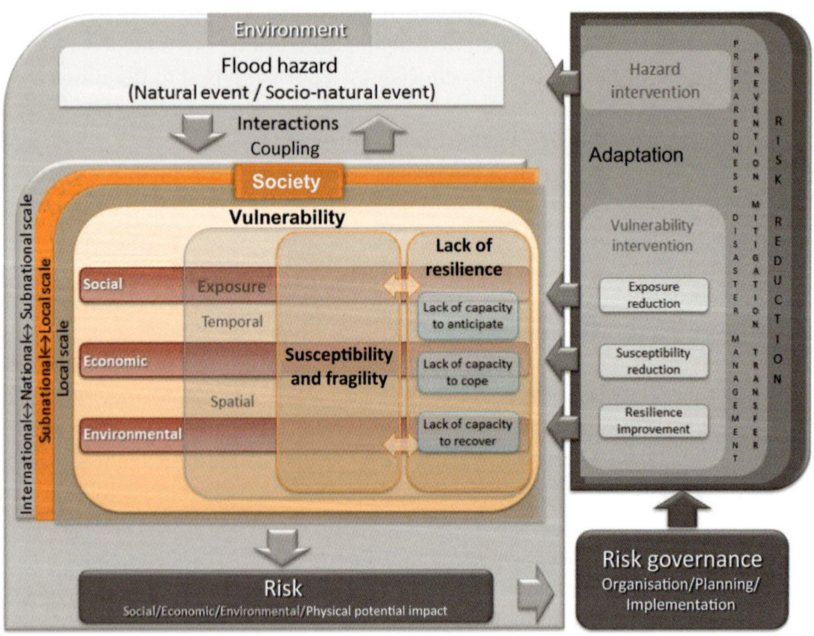

FIGURE 3.2 Adapted MOVE risk and vulnerability framework—causal factors of vulnerability assessed in the Salzach case study area are highlighted. Exposure was excluded from this assessment. *After Birkmann et al. (2013).*

specifically the vulnerability domain of the framework and its social, economic, and environmental dimensions individually, thus avoiding the challenge to identify an overall vulnerability index across several dimensions. The choice of the catchment scale level is associated to different policies implemented at this specific scale level (such as the water framework directive and the floods directive of the European Union, as well as national laws in Austria) and should therefore support decision makers at catchment level within their assessment of vulnerability and the appropriate and relevant identification of place-based intervention measures.

A joint workshop was carried out to critically reflect on the MOVE framework with different stakeholder groups on July 29th, 2010. This comprised representatives from flood protection, early warning, climate change adaptation, and the spatial planning departments of the Government of Salzburg, as well as representatives of the Meteorological Service of Austria. The feedback from the stakeholders regarding the generic framework was positive, as they were able to identify themselves and their activities within the framework. The social and economic dimension was seen as the most important one to be assessed within a vulnerability assessment and was therefore chosen next to the environmental and physical dimensions. As the framework and a holistic approach toward risk and vulnerability builds on a complexity of different understandings from disciplines related to the natural and social sciences, it is important to put emphasis on the appropriate presentation and communication of the framework. For instance, an animated version of the framework was used, which provided a stepwise reasoning of the different components, and included various examples which improved the understanding of the framework. An important component of the framework comprises the different intervention options in the risk governance component. This is an innovative and essential part of the MOVE framework, as a key objective of a vulnerability assessment becomes evident: the identification of relevant intervention options to reduce overall risk, while focusing on exposure, susceptibility reduction, and the improvement of resilience (see also Birkmann et al., 2013).

Based on the feedback gained from stakeholders and drawing on experiences made during the process of identifying relevant and salient indicators, the initial MOVE framework was slightly adapted in the following manner: (1) the assessed social, economic, and environmental dimensions were spanned over the subdomains of susceptibility and lack of resilience; and (2) the resilience domain has been relabeled to "lack of resilience" to be coherent with the negative association toward vulnerability. This allowed the specific and appropriate identification of indicators for each dimension not only for the susceptibility but also for the lack of resilience domain. Additionally, vulnerability has been assessed in a first instance, independent from the exposure domain. This is specifically argued for, as we aimed to assess the general potentiality of harm, independent from a hazard delineation, which is often based on certain model assumptions with related uncertainties. Combining vulnerability with hazard zones (frequency and extent indications) is considered as a next step toward the analysis of risk comprising hazard and vulnerability assessments. As such, vulnerability is independent from the frequency and extent of flood zones, and highlights potential vulnerable areas based on their predispositions and generic characteristics.

As mentioned above, the objective was to model homogenous regions of vulnerability independent of administrative boundaries. Being independent from such arbitrary and a priori geographies, we aimed to represent vulnerability place-based. This provides (1) an integrative and at the same time decomposable assessment of vulnerability and (2) the opportunity for place-specific (e.g., where are hot spots?) identification of appropriate intervention measures (what to address? and where?). The approach chosen builds on the concept of "geons" which has been defined by Lang et al., (2008) "as spatial units that are homogenous in terms of varying space-time phenomena under policy concern". More specifically, Lang et al., (under review) define geons as (derived from Greek gē (Γῆ)=land, earth, and the suffix-on=something being) a "specific type of region, semi-automatically delineated with expert knowledge incorporated, scaled and of uniform response to a phenomenon under space-related policy concern". Vulnerability units or regions are a subtype of such geons, labeled "integrated geons". Integrated geons address multidimensional and latent phenomena (=not directly measurable, such as vulnerability), and are operationalized by setting up a hierarchical conceptual model that, at the same time, enables a conceptual decomposition of the phenomenon, which is ultimately represented by a set of individual indicators (Lang et al., 2014). The approach builds on a concept to regionalize geospatial data layers and assign an integrated value (e.g., an index) to the derived regions.

The initial methodology to model such "integrated geons" or spatial vulnerability units was developed by Kienberger et al. (2009) in a similar vulnerability assessment in the Salzach catchment. However, this approach was built on an earlier IPCC definition of vulnerability and targeted solely the socioeconomic dimension. The developed approach has already been applied to a variety of application domains such as cumulative climate change impact in Western Africa (Hagenlocher et al., 2014) or vulnerability to floods in a rural setting in Mozambique (Kienberger, 2012).

An expert-based approach was selected by considering the fact that vulnerability is not directly measurable, due to its complex dimension and social construction. In order to model the spatial distribution of a complex phenomenon, established methodologies such as Multicriteria Decision Analysis, Delphi exercises, and the above mentioned regionalization approaches to derive homogenous vulnerability regions were integrated (Figure 3.3).

At the initial stage of the workflow, the concept of vulnerability is addressed, which in this case builds on the adapted MOVE risk and vulnerability framework. An essential step is the identification of indicators and available datasets, the weighting of the different indicators, and its aggregation to subindices. As mentioned above, expert consultations took place to develop an indicator framework which helped to identify different causes and triggers of vulnerability to

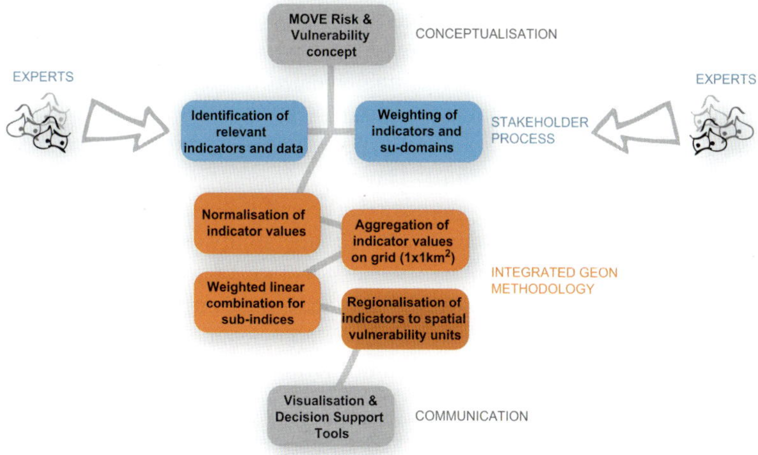

FIGURE 3.3 Applied workflow to model homogenous vulnerability units applying methods of multicriteria assessment as well as regionalization approaches *Based on Kienberger et al. (2009), Kienberger (2012).*

floods in the Salzach catchment, and to finally identify a set of suitable indicators. Such a stakeholder process is seen as important—next to a sound analysis of literature to justify the selection of indicators from a more scientific point of view. As the final list of appropriate indicators identified from expert choice, literature and community perspective, five key experts were asked to weight the indicators through budget allocation methods. Data availability has been an advantage in the Salzach catchment, as important key datasets are available in the spatial data infrastructure established at the Government of Salzburg, as well as the availability of detailed raster-based census data for Austria, which specifically helped to apply the regionalization approach. The data used range from infrastructure data to different socioeconomic parameters such as the size of companies, means of subsistence, age, and workforce in economy sectors, origin and education level of the population. They originated from the census survey in 2001 and are not only provided on the basis of different administrative units but also for a standardized grid with different available spatial resolutions (for details see Wonka, 2006).

To allow comparability, data were normalized through a linear min–max normalization within an 8 bit scale range (values with 0–255). The modeling scale level was based on the standardized grid with a regular cell size of 1000 m. In specific cases, it was required to develop subindices to provide the opportunity to weight also sub-indicators. For instance, the indicator "population age" was aggregated by judging and therefore weighting different age classes according to their social vulnerability (see below). The methodology to develop such subindices builds on weighted sum approaches, used in multi-criteria analysis (MCA, e.g., see Malczewski, 2000; see also Figure 3.3).

In the next step, the modeling of homogenous vulnerability regions for each dimension (as integrated geons) was implemented through the application of a multiresolution segmentation/regionalization approach (Baatz and Schäpe, 2000) in combination with a smoothing algorithm (polynomial approximation with exponential kernel algorithm). Therefore, regions of vulnerability have been delineated that share a commonality in regard to their underlying indicator values as well as a spatial constraint. A vulnerability index was calculated through a weighted vector magnitude (the length of the vector for each region) in the multidimensional indicator space. Final index values were then normalized within a scale range of 0–1, and visualized by applying a centile classification. Further details on the processes outlining normalization, final segmentation and smoothing of outlines have been published in Kienberger et al. (2009), Kienberger (2012), (Hagenlocher et al., 2014) as well as Lang et al., (2014).

3.3 INDICATORS FOR MULTIPLE VULNERABILITY DIMENSIONS

Following the conceptual MOVE framework, the Salzach case study undertook a vulnerability assessment integrating the causal factors, subdomains of susceptibility, and lack of resilience. The dimensions addressed comprise social, economic, and environmental characteristics separately. The assessment focused on each dimension independently, as it is difficult to compare and aggregate social with physical vulnerability factors for instance as they remain very much specific to each dimension, which makes it difficult to integrate those into an overall vulnerability index. Additionally, the dimension-specific approach helps to address "sectors" of decision making as proposed by Villagran (2006).

According to the MOVE framework (Birkmann et al., 2013), the three dimensions addressed in the Salzach catchment case study can be defined as follows:

- Social dimension: propensity for human well-being to be damaged by disruption to individual (mental and physical health) and collective (health, education services, etc.) social systems and their characteristics (e.g., gender, marginalization of social groups).
- Economic dimension: propensity for loss of economic value from damage to physical assets and/or disruption of productive capacity.
- Environmental dimension: potential for damage to all ecological and biophysical systems and their different functions. This includes particular ecosystem functions and environmental services (see, e.g., Renaud, 2006) but excludes cultural values that might be attributed.

Moldan and Dahl (2007) conclude by discussing the relevance of indicators and indices for policy, that indicators would be at the edge of science and politics and that outlining the issues in a policy relevant way is essential and requires a participatory process between users and scientists. They highlight that indicators need to be credible (scientifically valid), legitimate in the eyes of stakeholders, and salient and relevant to decision makers. These criteria also guided the selection process of the indicators as outlined above. Additionally, one of the central issues, especially within the assessment of vulnerability, is how we can integrate the various indicators toward a joint vulnerability measure and how they can be monitored over time, in order to identify trends and thresholds that can be useful for decision makers (Kienberger, 2012).

As an initial step, potential vulnerability indicators were identified from scientific publications, during the first phase of the project. In the second step, each single and composite indicator for the causal factors of the dimensions and capacities mentioned above were discussed through several sessions also with stakeholders in order to refine the list, considering expert and local knowledge. The main criteria to select the indicators were: relevance for the case study area, relevance to the particular hazard and its vulnerability, and—realistically—data availability, as it is depicted in Figure 3.4. Having a refined and final list expanded by a detailed literature review, the last step was to allocate weights for

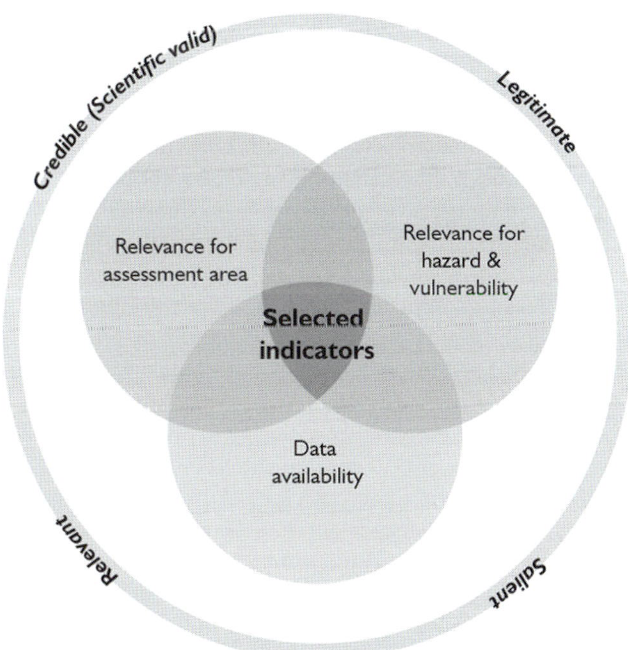

FIGURE 3.4 Criteria to indicator selection. From the universal set of vulnerability indicators, the indicators considered in the Salzach case study are the result of the intersection between the criteria: relevance for the case study area and the particular hazard and vulnerability, and data availability, based on general indicator selection criteria mentioned in the outer sphere.

further aggregation to each single and subindex integrating expert knowledge through the application of budget allocation methods.

In the following, we discuss key indicators for different dimensions, especially its relevance for the case study, and provide a hierarchical visualization of the indicator framework reflecting the applied weightings as tree maps (Shneiderman, 1992).

3.3.1 Social Dimension

The social dimension is considered by the MOVE framework as the propensity for human well-being to be damaged by disruption to individual and collective social systems and their characteristics. This dimension considers special groups of population, who are differentiated on accounts of their age, employment conditions, education level, gender, or origin.

Historically, the social dimension seems to be the most prominent one in the context of vulnerability assessment. It often aims to address the root causes and predispositions of societies to be affected by a hazard. Despite discussions on whether or not the social dimension can be assessed quantitatively or qualitatively, we aimed to provide a quantitative measure of social vulnerability, being at the same time aware that qualitative assessments may play a crucial and additional role for gathering a complete picture. Difficulties arise from the lack of empirical evidence of causal factors/"indicators" and often the lack of detailed data. However, the Salzach catchment provides an opportunity to develop enhanced methodologies, as detailed census results are available in a disaggregated manner—as mentioned above—on a regular grid level, and very detailed within their attributive resolution. Therefore, this provides opportunities to move toward place-based social vulnerability assessments not biased and influenced by arbitrarily built administrative boundaries (e.g., Cutter et al., 2003).

The social dimension (Figure 3.5) considers specific groups of vulnerable populations such as elders and children, due to the weakness of their physical and sometimes mental conditions. The most prominent application of several social vulnerability indicators is provided by Cutter et al. (2003). Additionally, Fekete (2009) applied for instance age-specific indicators for social vulnerability in the context of river floods in Germany. Extremes in the age spectrum may increase social vulnerability requiring to evaluate the vulnerability of those groups, in order to be able to develop policies to improve their conditions. Importantly, they are also considered as a special group in a preimpact recovery plan, to ensure the suitable management of their requirements.

Interestingly, experts weighted the different age groups quite equally (see Figure 3.5). This may be built on the assumption that social structures are believed to be strong in Austria. However, next to the employment sector and the availability of early warning systems in the resilience domain, age is one of three most critical factors in the social dimension.

Being employed in different sectors may provide different social vulnerabilities as sectors may be differentially affected by floods. For instance, farmers

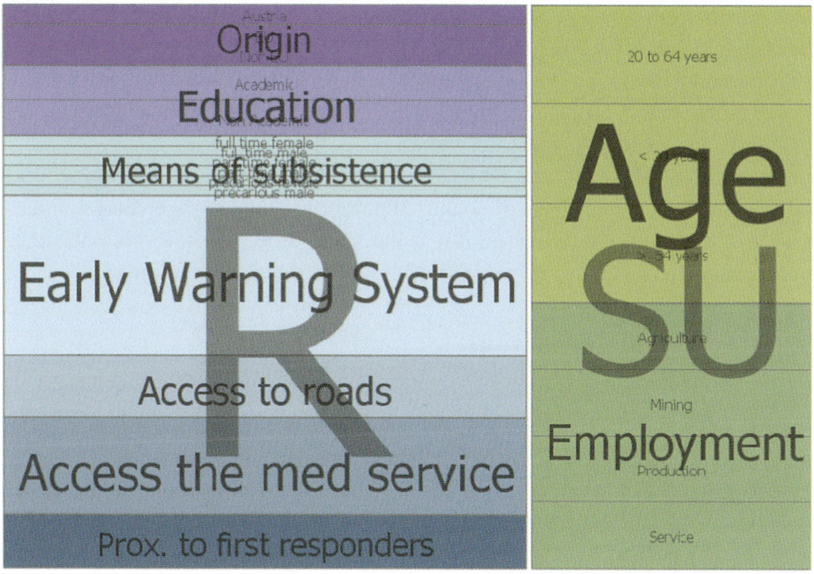

FIGURE 3.5 Lack of resilience (R) and susceptibility indicators (SU) for the social dimension visualized as a tree map. Tree maps are hierarchical representations. The size of the areas reflects the weights given to each indicator (e.g., different age classes) and subindices (e.g., overall age index).

being affected by a flood may lose their harvest, as also industry can be severely disrupted. Again those different sectors have been weighted equally, which may be due to available insurances as well as the "equality" of being employed in each sector. Age structure and sector of employment comprise the two most critical factors of social vulnerability in the applied case study according to the experts' weights. Less weight is given to the means of subsistence, which reflects the level of employment (full, part-time, precarious per gender). Again, this seems to be somehow contradictory to typical social vulnerability indicators, and highlights the assumption of the experts that those factors may play a minor role in this case study region.

Several authors have discussed the issue of immigrants and their related social vulnerability. Cutter et al. (2003) include the language and cultural barriers in their social vulnerability index because they affect the access that immigrants have to some services in the predisaster phase and funding in the postdisaster phase. Fekete (2009) within his validation of the social vulnerability index for Germany included also new residents and he provided a gender consideration when he included foreign females as one of the indicators for social vulnerability. However, stakeholders assigned the least weight and merely assigned equal weights to different "origin of population" classes. Subjectively, this may be due to sensitivity issues of marginalizing certain social classes.

In regard to education, Cutter et al. (2003) state that a low level of education reduces the comprehension ability to understand warning information and information during recovery. The educated population has more access to information, than less well educated. Again, this indicator group received less attention and education groups have not been differentiated internally.

Overall, much emphasis is given to the lack of resilience domain, and response and recovery indicators such as early warning systems, access to roads, medical services, and first responders. As such, the Salzach case study reflects some key indicators of social vulnerability, namely age classes and the sector of employment, while giving at the same time high relevance to operational response and recovery activities affecting social vulnerability.

3.3.2 Economic Dimension

The economic dimension deals with the potential financial damage and/or disruption of the productive capacity. As loss of life is fortunately very much limited in flood events in the Salzach catchment, most of direct and indirect damage originates from disruption of the productive capacity. Cutter et al. (2003) include transportation infrastructure in their social vulnerability index taking into account that the loss of this infrastructure is a difficult financial burden on communities. The transport network and its subclasses such as highways, primary roads, railways, and secondary roads received also the highest attention through the weighting exercise (Figure 3.6). This reflects the importance of transportation in general, but also specifically the importance of transportation networks

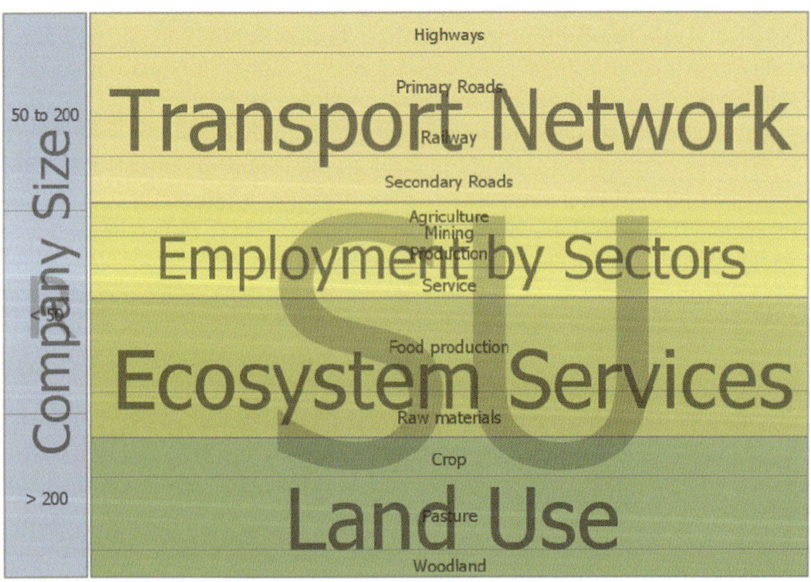

FIGURE 3.6 Weighted indicators and subindices for the economic dimension, grouped into the lack of resilience (R) and susceptibility domain (SU).

in the Salzach catchment comprising important *trans*-alpine and east-west transportation corridors in Central Europe, which were severely affected during the June 2013 events.

Additionally, ecosystem services play a particular role in the catchment. Especially, those services providing food production (mainly dairy farming) as well as raw materials (forests). Additionally, specific land use classes such as pasture but also crops and woodland (less weight) received attention by the different experts.

Overall, less weight is given to the employment by different sectors (same indicator as in the social dimension), as well as the company size. According to the expert weighting company size is not considered to substantially effect levels of vulnerability (by numbers of employees).

In total much weight is given to the susceptibility domain, which is due to the lack of indicators on the resilience side, but also due to the importance of certain indicators representing predispositions in the susceptibility domain. It is furthermore important to note that the economic dimension does not represent any monetary values, but aims to reflect the vulnerability of the "economic" system in a spatial manner. Often the social and economic dimensions are being combined to an overall socioeconomic metric. However, despite the challenge of identifying valid indicators for each dimension separately, both dimensions have been attributed by the stakeholders as the most relevant ones for the case study region. Specifically, weight is given to the economic dimension, as experienced during the 2013 events; as loss of life or people being injured occurs less as compared to direct or indirect impacts on the economy.

3.3.3 Environmental Dimension

The environmental dimension deals with the fragility of ecological and biophysical systems, and their different functions, under a hazardous condition, to suffer damage and deterioration. Assessing the environmental dimension is often neglected, due to the lack of awareness on causes of environmental degradation for flood impacts as well often being "overruled" by socioeconomic factors, comprising the heart of current assessments. However, the environmental dimension needs to receive further attention, as it may provide specific insights also for mid- to long-term effects, and especially in the context of climate change.

Rivers, lakes, and wetlands denominated fresh water ecosystems are extremely rich in species; however, they are also the most altered and threatened ecosystems in the world. Essential characteristics as the natural flow regime, and the longitudinal and lateral connectivity of rivers, to sustain the biophysical and ecological processes required for life in fresh waters, are disrupted by dams and their reservoirs in fragmented rivers. The fragmentation and hence the loss of ecosystems processes not only affect ecosystem and species but also

FIGURE 3.7 Weighted indicators and subindices for the environmental dimension.

humans. The loss of floodplain inundation patterns affect human communities due to changed run-off patterns as well as settlements/infrastructure on "former" floodplains (Figure 3.7).

Therefore, we aimed to represent the environmental dimension through three available subindices, namely, hemeroby as a proxy measuring the "naturalness" of the landscape, degradation of rivers, and forest damages of landscapes as well as ecosystem services, on disturbance regulation, soil retention, and water regulation.

According to Kowarik (1988), hemeroby is "a measure for man's cultivation influence on ecosystems, whereas assessment of hemeroby level is carried out based on the extent of effects of those anthropogenic influences which affect the system's development towards a final state". The hemeroby scale relies on the vegetation coverage and the properties of habitats. Including this indicator in the environmental vulnerability assessment builds on the assumption, the more a landscape is kept natural, the less severe the impacts from peak flows will be. As the natural environment in central Europe is already kept to a minimum, this may be an important factor to consider while adapting also for future climate change. In this context, a dataset reflecting hemeroby in Austria was applied (see Rüdisser et al., 2012). Hemeroby was rated as the most important indicator in this domain.

The second most important indicator is a subindex reflecting degraded landscapes. One important factor in this context is the regulation of Salzach river, which is a cause for changed and increased flood impacts. Currently efforts are being undertaken to "reconstruct" the natural river flow, which is expected to reduce peaks in run-off as well as flood impacts in general. An equally important

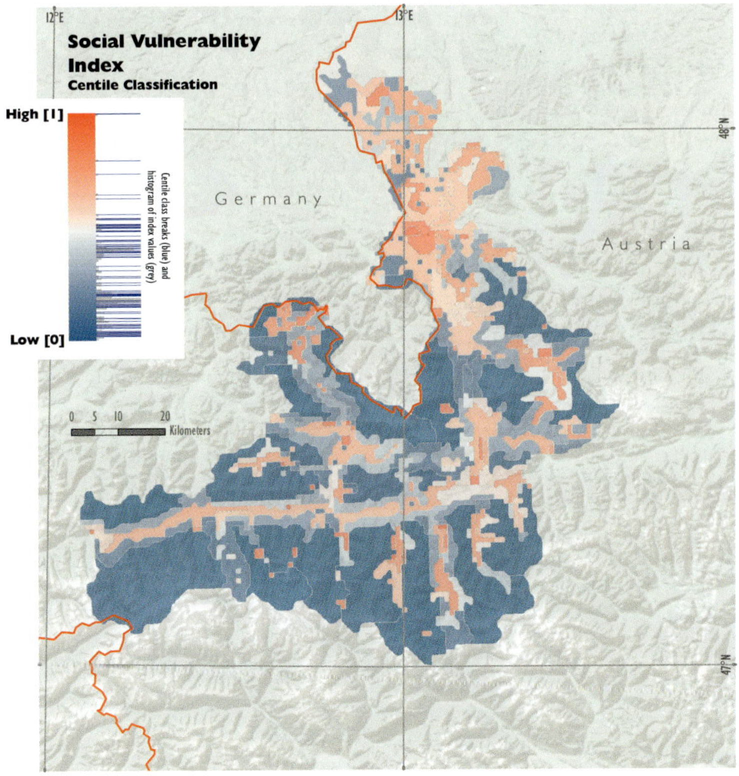

FIGURE 3.8 Social vulnerability.

factor is the state of forests, as they comprise the capacity to "store" water during peak precipitation events. In the past, forests have been damaged either by bark beetle or due to mono cultivation, or through past storm events impacting local water regimes. Therefore, it is important to consider such degradation factors also in the light of regular monitoring activities.

As already incorporated in the economic dimension, ecosystem services play a significant role in the environmental dimension as well. In general, the approach followed here builds on Costanza et al. (1997) where different ecosystem services are identified and valued for different land cover classes. As such we applied the values as a relative measure of importance for relevant services in specific domains. In the context of flood events and the environmental dimension, the services on disturbance regulation, soil retention, and water regulation play a critical role.

3.4 RESULTS

Vulnerability maps for the social, economic, and environmental dimension, based on the geon approach and integrating expert weights are shown in Figures 3.8–3.10. The vulnerability index is derived as the weighted vector magnitude in the multi-indicator space. Values have been reclassified from zero (0) to one (1),

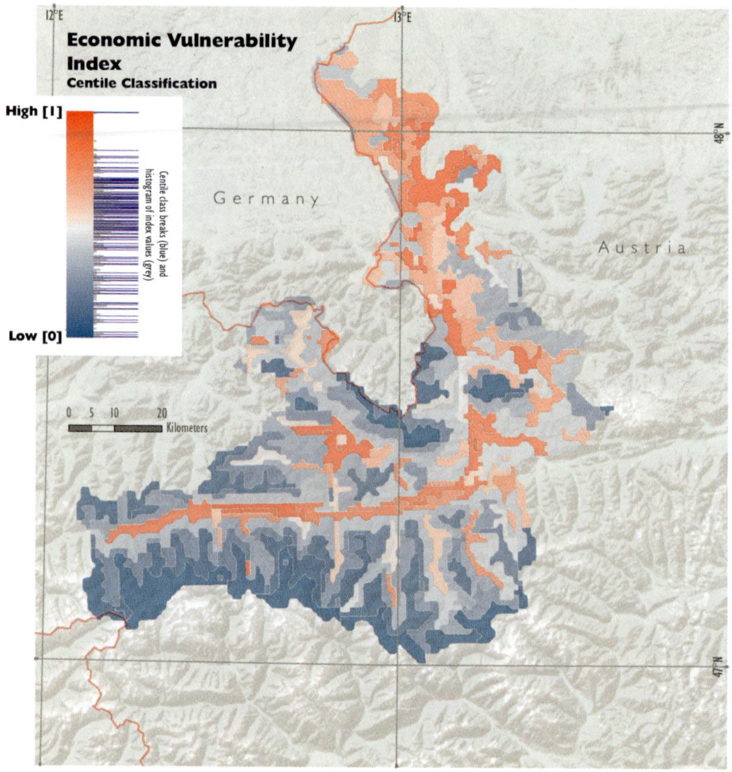

FIGURE 3.9 Economic vulnerability.

whereas one represents the highest, relative (!) vulnerability value. Results are classified as a centile classification and visualized using a continuous blue-to-red color scheme. The high number of classes and its semicontinuous display emphasizes the independency of class biases. For transparency reasons, the centile class breaks and the associated histogram of the index values is presented in the maps.

The social dimension (Figure 3.8) shows specific hot spots along the major river valleys of the Salzach catchment, as well as in city centers. It is interesting to observe that the city of Salzburg itself serves as a hotspot with an accentuation toward the north-western part of the city. Such hotspots are characterized not only through differentiation in the demographic values, such as age, means of subsistence as well as education, but also in certain areas with lack of access to response and recovery services. Overall it is interesting to observe that the geons follow the major settlement characteristics which underline their notion of being independent from administrative boundaries.

The results for the economic vulnerability assessment in Figure 3.9 show several areas with a medium degree of vulnerability along the Salzach river and some of its tributaries. These regions are surrounded by other areas with a medium degree of vulnerability along all the tributaries of the Salzach river and the rest of the area is considered with a low degree of vulnerability. The point with the

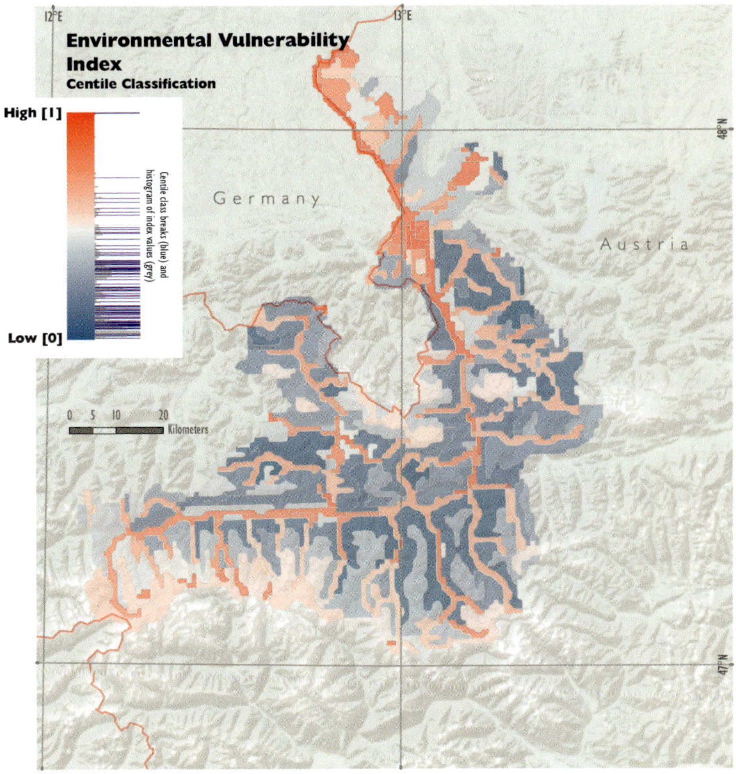

FIGURE 3.10 Environmental vulnerability.

highest degree of vulnerability in the north-western part of the city of Salzburg accounts for the presence of employment sources, the lack of ecosystem services, and being an important node in the transport network in the country.

Within the vulnerability assessment, the environmental dimension (Figure 3.10), shows highest levels concentrated around the city of Salzburg and north and south of it, where a lack of regulating ecosystem services exists as well the "naturalness" of landscape has been lost. These regions are followed by an area with a high-medium degree of vulnerability in the north which is due to strong river fragmentation currently being rehabilitated. Some other regions with medium degree of vulnerability located mainly along some of the tributaries of the Salzach are impacted by different environmental degradation processes.

3.5 DISCUSSION

One added value of the results obtained relies on the possibility to use the geon concept and vulnerability regions to assess the vulnerability to floods combined with mixed methods of stakeholder's involvement, multicriteria assessment and regionalization. The applied method is based on the geon concept (Lang et al., 2008) and the associated methodology (Kienberger et al., 2009) as the framework aims to represent vulnerability independently from administrative units and therefore reduce unit-related biases as modifiable areal unit problem and related effect of "ecological fallacy". Additionally, and this as a core objective, it enables the identification of place-specific intervention measures. The methodology proposed was implemented in the Salzach river. It allows transforming continuous spatial information into discrete objects, in order to monitor changes according to different hazards in any other area. Further research is needed to integrate methods from composite indicator construction and multicriteria assessment to provide an evaluation of the statistical soundness of the chosen indicator set adapted to the regionalization approach applied here.

The approach can be integrated to methodologies for risk assessment or it can be applied in further research focusing more on some specific causal factors or dimensions of vulnerability. In the specific case of flood hazards, the methodology proposed is an attempt to generate results that support the requirements of the EC Flood Directive in regard to water policies for developing river basin management and flood risk management plans, empower community actions, having informed technical, financial, and political decisions and develop adapted prevention, protection, and preparedness measures.

Next to the identification of hot spot areas, the geon approach facilitates the identification of place-based interventions. As shown in Figure 3.11, regions can be queried and the relative share of the underlying indicators can be visualized. For instance, the two selected regions not only differ in their vulnerability value but also in the relative contribution of vulnerability factors. This should enable stakeholders to identify priority regions (hot spots) but also identify appropriate intervention options which may differ from one region to the next. For instance in Figure 3.11, the access to hospitals is an issue in region 2, whereas this factor can be neglected in region 1.

There are several challenges and difficulties related to data availability, integration of exposure, susceptibility, and lack of resilience toward vulnerability and difficulty to identify "lack of resilience" indicators.

In the discussions with stakeholders, several important indicators were mentioned, mainly in the resilience domain such as the importance of social networks and communication and media coverage, as well as the risk perception of the population. However, proxies and available data are insufficient to reflect them in an appropriate manner. The selection of indicators may have to be very place-specific as the characteristics of vulnerability may differ.

3.6 CONCLUSIONS AND RECOMMENDATIONS

The most vulnerable regions in the three dimensions are located along the Salzach river and its tributaries. However, the most vulnerable region in all the dimensions is the city of Salzburg and its surroundings. These results are due

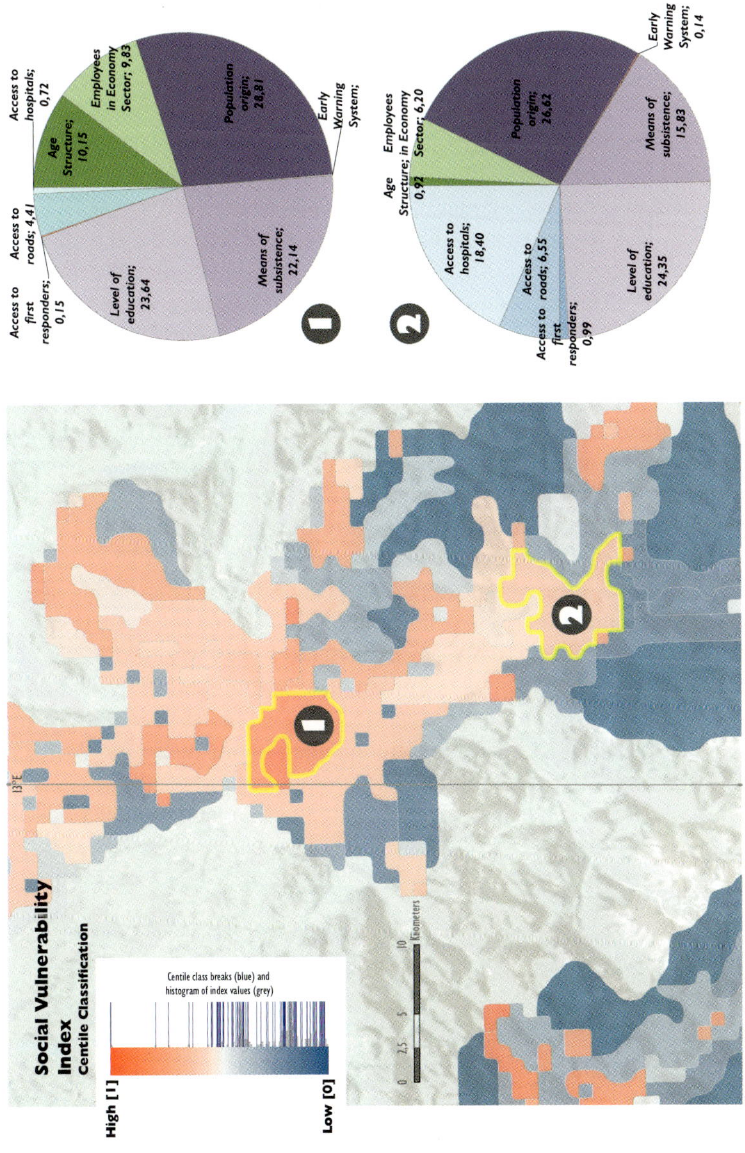

FIGURE 3.11 Identification of two selected vulnerability regions and the related visualization of the relative share of the underlying indicators per region.

to the density of the built up area, a big concentration of historic buildings and wide-spread infrastructure. Therefore, the highest vulnerability degree in regard to the social dimension is concentrated in the city of Salzburg as one of the largest settlements located along the Salzach river. The economic dimension has its hot spot also in this urban region due to the presence of employment sources, and the city being an important node in the transport network in the country. The environmental dimension shows the highest degree of vulnerability in the stretch of the Salzach river from the city to the north, associated to river fragmentation.

The MOVE framework successfully enabled us to identify relevant indicators for each of the investigated dimension. As the framework was accepted by the stakeholders, it served as a common ground for detailed discussions on risk reduction measures. Rather than finding problems to understand a specific part of the framework, challenges have been identified in understanding the relationship between the different components as empirical evidence of contributing factors is very limited. This also underlines and acknowledges the complexity and "messiness" of how vulnerable conditions actually come about. Therefore, a vulnerability assessment can be seen as an approximation toward reality and a stimulus for critical reflection. However, such a spatial approach should enable us to be more place and content-specific than roaming within rather general recommendations. Of course inherent uncertainties and assumptions need to be critically communicated with any assessment results.

Future attention should be given to the dynamic and complex characteristic influenced by various factors/indicators, processes and drivers operating at different spatial and temporal scales, which are interlinked and mutually reinforcing each other. However, this provides the ground for a key objective of any vulnerability assessment: enabling a continuous monitoring in space and time, which allows the identification of critical trends as well as the monitoring and evaluation (and impact assessment) of different intervention measures. With this approach, we move one step forward toward providing policy relevant information in the context of disaster risk reduction and climate change adaptation.

ACKNOWLEDGMENTS

The authors would like to thank Judith Huber, Gerald Reischenböck, Michael Hagenlocher, Lena Pernkopf and Raphael Spiekermann for their support.

REFERENCES

Baatz, M., Schäpe, A., 2000. Multiresolution segmentation: an optimization approach for high quality multiscale image segmentation. In: Strobl, J., Blaschke, T., Griesebner, G. (Eds.), Proceedings of the Angewandte Geographische Informationsverarbeitung XII, pp. 12–23.

Birkmann, J., Cardona, O.A., Carreño, L., Barbat, A., Pelling, M., Schneiderbauer, S., Kienberger, S., Keiler, M., Zeil, P., Welle, T., 2013. Framing vulnerability, risk and societal responses: the MOVE framework. Natural Hazards. 67 (2), 193–211. http://dx.doi.org/10.1007/s11069-013-0558-5.

Costanza, R., D'arge, R., De Groot, R., Farber, S., Grasso, M., Hannon, B., Limburg, K., Naeem, S., O'Neill, R., Paruelo, J., Raskin, R.G., Sutton, P., Van den Belt, M., 1997. The value of the world's ecosystem services and natural capital. Nature 387, 253–260.

Cutter, S.L., Boruff, B.J., Shirley, W.L., 2003. Social vulnerability to environmental hazards. Social Science Quarterly 84 (2), 242–261.

Fekete, A., 2009. Validation of a social vulnerability index in context to river-floods in Germany. Natural Hazards and Earth System Sciences 9, 393–403.

Hagenlocher, M., Lang, S., Hölbling, D., Tiede, D., Kienberger, S., 2014. Modeling hotspots of climate change in the Sahel using object-based regionalization of multidimensional gridded datasets. IEEE Journal of Selected Topics in Applied Earth Observations and Remote Sensing (JSTARS). 7 (1), 229–234. http://dx.doi.org/10.1109/JSTARS.2013.2259579.

Kienberger, S., 2012. Spatial modelling of social and economic vulnerability to floods at the district level in Búzi, Mozambique. Natural Hazards. 64 (3), 2001–2019. http://dx.doi.org/10.1007/s11069-012-0174-9.

Kienberger, S., Lang, S., Zeil, P., 2009. Spatial vulnerability units – expert-based spatial modelling of socio-economic vulnerability in the Salzach catchment, Austria. Natural Hazards and Earth System Sciences 9 (3), 767–778.

Kowarik, I., 1988. Zum menschlichen Einfluss auf Flora und Vegetation: Theoretische Konzepte und ein Quantifizierungsansatz am Beispiel von Berlin (West). Landschaftsentwicklung und Umweltforschung 56, 1–280.

Lang, S., Zeil, P., Kienberger, S., Tiede, D., 2008. Geons – policy-relevant geo-objects for monitoring high-level indicators. In: Car, A., Griesebner, G., Strobl, J. (Eds.), Geospatial Crossroads @ GI_Forum. Proceedings of the Second Geoinformatics Forum Salzburg. Wichmannn, Heidelberg, pp. 180–185.

Lang, S., Kienberger, S., Tiede, D., Hagenlocher, M., Pernkopf, L., 2014. Geons - domain-specific regionalization of space. Cartography and Geographic Information Science 41 (3), 214–226.

Malczewski, J., 2000. On the use of weighted linear combination method in GIS: common and best practice approaches. Transactions in GIS 4 (1), 5–22. http://dx.doi.org/10.1111/1467-9671.00035.

Moldan, B., Dahl, AL., 2007. Challenges to sustainability indicators. In: Hak, T., Moldan, B., Dahl, AL. (Eds.), Sustainability Indicators. A Scientific Assessment. Island Press, Washington, DC, USA, pp. 1–26.

Renaud, F.G., 2006. Environmental components of vulnerability. In: Birkmann, J. (Ed.), Measuring Vulnerability to Natural Hazards: Towards Disaster Resilient Societies. United Nations University Press, Tokyo, pp. 117–127.

Rüdisser, J., Tasser, E., Tappeiner, U., 2012. Distance to nature – a new biodiversity relevant environmental indicator set at the landscape level. Ecological Indicators 15, 208–216.

Schmid, D., Lederer, I., Much, P., Pichler, A.-M., Allerberger, F., 2005. Outbreak of norovirus infection associated with contaminated flood water, Salzburg, 2005. Eurosurveillance 10 (24).

Shneiderman, B., 1992. Tree visualization with tree-maps: 2-d space-filling approach. ACM Transactions on Graphics 11 (1), 92–99.

Ulbrich, U., Brücher, T., Fink, A.H., Leckebusch, G.C., Krüger, A., Pinto, J.G., 2006. The central European floods of August 2002: part 1 – rainfall periods and flood development. Weather 58 (10), 7.

Villagrán, J.C.V., 2006. Vulnerability: A Conceptual and Methodological Review. UNU Institute for Environment and Human Security. Source Nr. 4.

Wonka, E., 2006. Regionalstatistik in Ö sterreich. In: Strobl, J. (Ed.), Von der Tabelle zu räumlicher Analyse und Visualisierung,I: Salzburger Geographische Arbeiten, Band 39. Salzburg und Wien.

Chapter 4

Vulnerability to Earthquakes and Floods of the Healthcare System in Florence, Italy

Roberto Miniati* and Christian Iasio†

*Department of Information Engineering, University of Florence, Florence, Italy, †Institute for Applied Remote Sensing, EURAC European Academy for Research, Bolzano, Italy

Chapter Outline

4.1 INTRODUCTION AND BACKGROUND

As hospitals represent the final point in the whole short and medium rescue chain, they are one of the most essential infrastructures to support victims and affected people during disasters, subsequently the operational continuity and the safety of patients and staff must be ensured. Health facilities are usually exposed to many hazards but earthquakes and floods represent the worst case scenario for hospital performance degradation. According to the international guidelines (PAHO, 2000; WHO Europe, 2006), earthquakes represent the most relevant risk for medical facilities as they are responsible for internal and external damages and for causing high medical need in the affected local population. Floods can also directly affect health structures and cause a high number of casualties and deaths. It represents the most common natural disaster and the broadest in terms of population and countries affected (Caballeros and México, 1994; OP, 2006; Federal

Assessment of Vulnerability to Natural Hazards. http://dx.doi.org/10.1016/B978-0-12-410528-7.00004-7

Emergency Management Agency, 2003). According to the World Disaster Report (International Federation of Red Cross 2002) they represent 35% of all disasters between 1992 and 2001.

The aim of this work is to systematically evaluate the health system's ability to respond to natural disasters in the Province of Florence (Italy). To do so, this paper analyzes the region's five most important hospitals and the ambulance service headquarters. As the study focused on the local level (province, municipalities, and some individual buildings), the stakeholders who were asked to participate were local institutional decision-makers, officials responsible for aspects of the health system, and functionaries of the local civil protection department (Dubois and Prade, 1998).

The seismic risk in Tuscany is moderately high, especially for the Apennine mountain area north of Florence (Mugello) where, as reported in the MOVE Project Handbook (2011), the last major seismic disaster in Tuscany was registered on the 29th of June 1919, with an estimated magnitude of 6.2 moment and damages at level X on the version of the Mercalli scale used at the time. Seventy towns were significantly affected and approximately 100 people were killed and 400 injured (Castenetto and Sebastiano 2004).

Similarly, flood risk affects many cities in Tuscany. Although Tuscany amounts to between 6% and 7% of the territory of Italy, 20% of major floods to take place in Italy during the last 60 years took place in this region. In total, during this timeframe, there were 10 major inundations. (http://it.wikipedia.org/wiki/Lista_di_alluvioni_e_inondazioni_in_Italia)."

One of the largest single floods of modern times occurred in Florence on 3rd of November 1966 and caused 34 deaths.

Figure 4.1 shows the geographical exposure of Florence's current hospital system at the levels reached by the 1966 floodwaters (intervals of 0–2 m, 0–4 m, 4–5 m) where the numbers indicate the Florentine medical facilities included in the study. The headquarters of the emergency ambulance service (addressed by the letter "a" in figure 4.1) and one 90-bed hospital would be well inside the flooded area.

4.2 MATERIAL AND METHODS

The methodology consisted of the following steps:

- Application of the MOVE general framework
- Development of a model for hospital vulnerability assessment
- Disaster risk evaluation in hospitals (performance degradation and economic loss)

4.2.1 Application of the MOVE General Framework

As the aim of the study was to assess the vulnerability to earthquakes and floods of the Florence hospital system, the MOVE general framework was applied

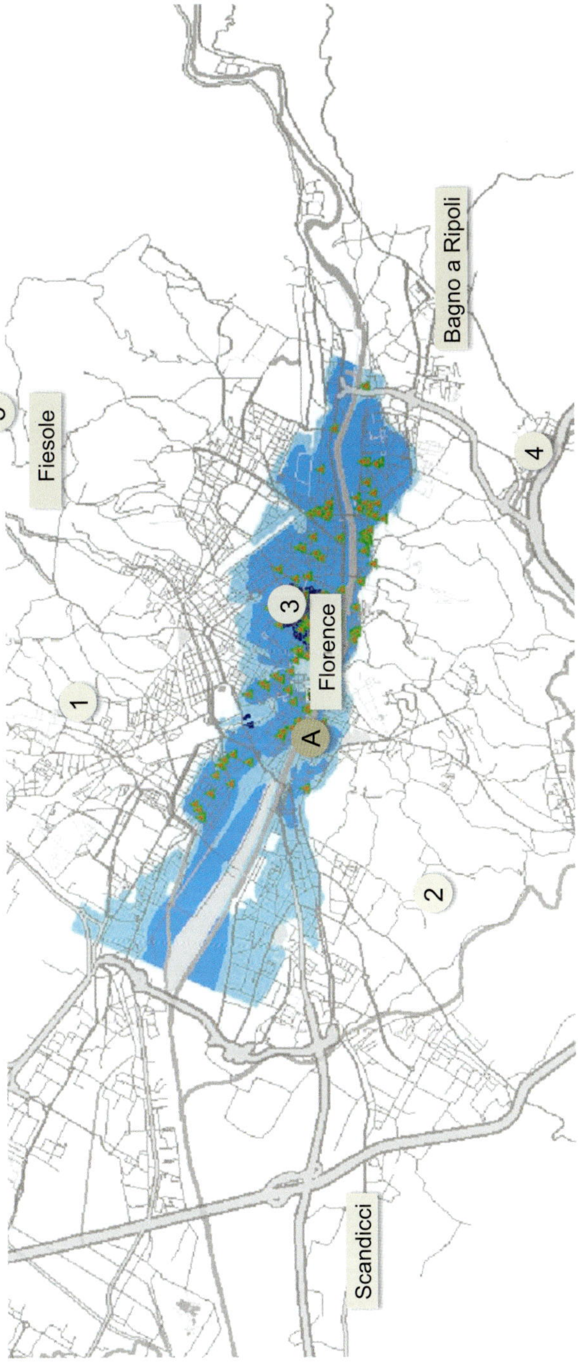

FIGURE 4.1 Exposure of the Florence hospital system to the 1966 flooding scenario.

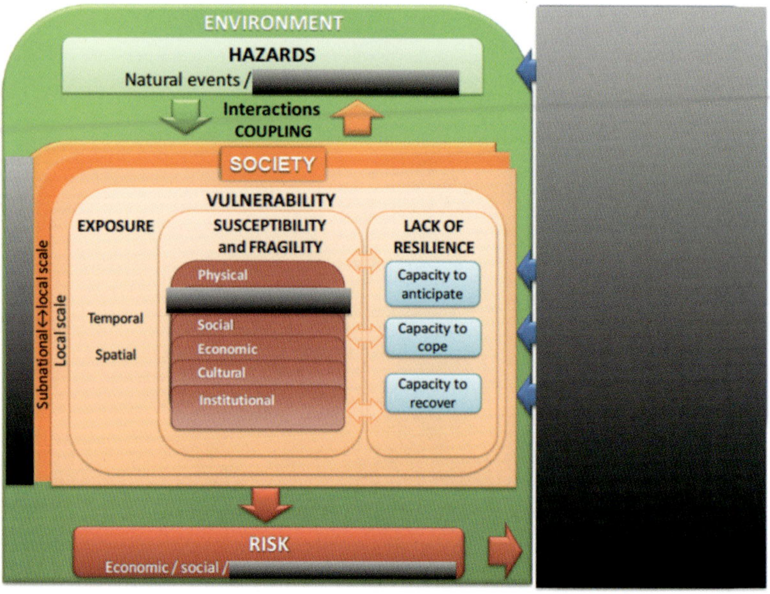

FIGURE 4.2 MOVE project general framework applied to Tuscany case study.

to the case study. The hospital's vulnerability assessment included both geographical and temporal exposures through the development of two scenarios representing a nighttime/holiday and a daytime/working-day scenario. Since hospitals are complex systems with performance levels dependent on different factors, a multidimensional analysis was used, including the evaluation of susceptibility, fragility, and estimation of lack of resilience. The risk methodology considers economic structural vulnerability from floods and social vulnerability from both earthquake and floods (Figure 4.2).

4.2.2 Hospital Vulnerability Assessment to Floods and Earthquakes

Several methodologies have been developed and applied for assessing vulnerability of hospitals, including several models from the complex systems theory or practical guidelines for rapid visual assessment (Setola, 2007; Barton et al., 2000; Bolona and Casalicchio, 2007; Commission of the European Communities, 2006; Panzieri et al., 2005; Setola and De Porcellinis, 2008). However, no single method currently exists or is widely endorsed for the vulnerability assessment of health structures. Some methodologies have made commendable progress toward encompassing the complexity of system organization (Setola, 2007); however, they have not performed reliably enough with real data. Others have worked better with real data (WHO Europe, 2006; PAHO, 2000); but they have considerable gaps in their description of the complexity within the hospital system. For this reason, an integrated methodology was proposed, based on a combination of the approaches: the theory of complex systems by using the Input–Output Inoperability Model of Leontief (Haimes and Jiang, 2001), and rapid seismic vulnerability assessment through field data collection by applying the World Health Organization evaluation forms (WHO Europe, 2006).

The Wassily Leontief model on the balance of economic systems forms the basis of the input–output model of inoperability (IIM Inoperability Input–Output Model) developed by Haimes and Jiang (2001) and largely used for the analysis of complex infrastructures. The model is able to describe the impact of an initial shock on the infrastructure and the propagation to the interconnected systems by considering how a specific failure can influence the functionality of the whole system. Three elements are necessary for the application of the Leontieff model: the definition of the elements composing the system, the mutual connections among hospital areas, and estimation of the inoperability vector resulting from the shock (Iadanza et al., 2009).

A list of hospital elements needed in the aftermath of a disaster is reported as follows (World Health Organization Year?):

A—Emergency department
B—Diagnostics
C—Surgical operation
D—Intensive Care Unit (ICU)
E—Hospital beds
F—Morgue
G—Laboratory
H—Crisis room
I—Accessibility and internal viability
L—Power network
M—Gas network
O—Back-up generator
P—Fire system
Q—Medical equipment

This proposed approach enables the construction of a knowledge-based system through the submission and compilation of semistructured questionnaires from experts estimating the interdependencies of the hospital system (Dubois and Prade, 1998; Kosko, 1994; Salski et al., 1996). In order to reduce the uncertainty associated with the data compiled, this methodology uses a fuzzy logic model which takes into consideration both the user's expertise and the level of confidence placed in the answer. The selected experts were asked to estimate the inoperability level of the elements represented as row in the matrix caused by the complete failure and stoppage of each element as columns. For instance, participants were asked to quantify the impact on each hospital area due to the complete shutdown of the emergency department by considering five inoperability levels (from "no effect" to "complete stoppage") and five confidence levels (from "strongly disagree" to "completely agree"). Each opinion was then weighted on the expertise level according to the position of the participant in the hospital knowledge hierarchy, the highest being the technical director (assigned with a weight of 1.0), followed by the medical director (0.75), while the lowest being the regular hospital engineers (0.5) and physicians (0.3) (Miniati and Iasio, 2012).

The internal connections between sections of the hospital system are defined as "Leontief matrix coefficients" and include the mutual interdependencies of functional impacts (e.g., element "*a*" is essential to the performance of element *b*, but element *a* is not influenced at all by the operational level of element "*b*"). The inoperability of the infrastructure can assume a value between 0.0 and 1.0, 0.1 corresponding to a perfect performance level (no degradation) to the complete ineffectiveness of the system (i.e., 100% inoperability), respectively. The model can be summarized as follows:

- x_j—overall risk of ineffectiveness of the *j*th interconnected infrastructure that can be caused by one or more failures
- r_i—*i*th resource to manage the risk of ineffectiveness of the specific critical infrastructure due to the general complexity of the system
- x_{kj}—the degree of subsequent ineffectiveness of the *k*th infrastructure due to one or more multiple failures caused by the *j*th interconnected infrastructure element
- a_{kj}—a weight representing the influence that the *j*th infrastructure has on the ineffectiveness of the interconnected *k*th infrastructure, due to the type and strength of their interconnection: its values form the interdependencies coefficient of matrix A.

Where in case of using the matrix form Eqn (4.1):

$$\begin{pmatrix} x_1 \\ \cdots \\ x_n \end{pmatrix} = \begin{pmatrix} a_{11} & \cdots & a_{1n} \\ \cdots & \vdots & \cdots \\ a_{n1} & \cdots & a_{nn} \end{pmatrix} \begin{pmatrix} x_1 \\ \cdots \\ x_n \end{pmatrix} + \begin{pmatrix} c_1 \\ \cdots \\ c_n \end{pmatrix} \tag{4.1}$$

and considering the synthetic form Eqn (4.2):

$$X = \mathbf{C}(I - \mathbf{A})^{-1} \tag{4.2}$$

We obtain the final form Eqn (4.3):

$$X = (I - A)^{-1} C = \left(\begin{bmatrix} 1 & 0 & 0 & 0 \\ 0 & 1 & 0 & 0 \\ 0 & 0 & 1 & 0 \\ 0 & 0 & 0 & 1 \end{bmatrix} - \begin{bmatrix} a_{00} & a_{01} & \cdots & a_{0n} \\ a_{10} & a_{11} & \cdots & a_{1n} \\ \cdots & \cdots & \cdots & \cdots \\ a_{n0} & a_{n1} & \cdots & a_{nn} \end{bmatrix} \right)^{-1} \begin{bmatrix} c_0 \\ c_1 \\ \cdots \\ c_n \end{bmatrix} \tag{4.3}$$

When the system is fully operating, each $x_k = 0$ for every *k* and $\Sigma x_k = 0$.

The last step in the implementation of the model was the definition of the input failures vector which provides the initial level of failures of each hospital infrastructure by considering a specific disaster scenario. In order to assess the initial impact or the level of inoperability, rapid vulnerability evaluation forms were used.

For the seismic scenario, the forms designed by the World Health Organization—Europe Office for rapid seismic vulnerability assessment forms (WHO-Europe, 2006)—were used. For each health facility, considering a fixed seismic intensity, it is possible to estimate the level of vulnerability (as high, moderate, or low) by considering the visual analysis of structural aspects (year and material of construction, elevation symmetry, etc.), nonstructural elements (anchoring technologies, basic lifelines, etc.), and organizational factors (staffing levels, presence of strategic medical areas, back-up of equipment, etc.). Each vulnerability level provided a numerical inoperability value in the failure vector, as follows: high = 0.3, moderate = 0.1; low = 0.03.

According to the flooding scenario, Figure 4.3 reports the rapid vulnerability assessment form used for the evaluation. The form was redesigned by taking

Rapid flood vulnerability assessment: H3

Flooding risk exposure: High

Type	Number of beds
Emergency dept.	20
Surgery	18
ICU	5
Sub ICU	3

Health functions	Underground	Ground level	First level and upper
ICU		X	
Emergency dept.		X	
Diagnostic		X	
Surgery			X
Urology			X
Pharmacy		X	
Sterilization			X
In-patient			X
Laboratory		X	
Blood bank		X	

Systems	Underground	Ground level	First level and upper
Medical gas central		X	
Power central	X		
Ups and back up generators	X		
Data and servers		X	

FIGURE 4.3 Example hospital evaluation form for rapid vulnerability assessment to floods.

into consideration the exposure of each hospital area to floods, their physical location within the health structure of medical areas, basic lifeline components, as well as medical equipment in the underground, ground, or upper levels.

4.2.3 Disaster Risk Evaluation in Hospitals

In order to estimate the disaster risk of hospitals, the performance degradation and the economic loss were analyzed. For the performance degradation, the Hospital Treatment Capacity (HTC) pre- and post-event estimates were used (De Boer, 1999). This index is defined as the number of patients per hour that the hospital is able to surgically treat. It depends on physical and organizational resources provided by the hospital especially for surgical activities. Equation (4.4) reports the HTC calculation defined as the number of fully operational surgical tables available per hour. Coefficient α (0–1) describes the organizational component, β (0–1) is the staff contribution, γ_1 the number of full operational surgical tables, $\gamma 2$ (0–1) is the surgery area operational level and t_m describes the standard duration for surgeries (Lupoi et al., 2008).

$$\text{HTC} = \alpha \times \beta \times (\gamma_1 \times \gamma_2) / t_m \, [\text{n}^\circ \text{ treated patients/h}] \quad (4.4)$$

The Intrinsic Security (IS) index (Miniati and Iasio, 2012) was created to evaluate the safety and the security of the most exposed people within the structure, provided by regular and ICU inpatients. For hospitals, safety and security should be highlighted for the high density of incapacitated or dependent people 24h per day. The new index "IS" is reported in Eqn (4.5):

$$\text{IS} = \alpha \times \beta \times [(\gamma_3 \times \Gamma_2) + (\gamma_4 \times \rho_2)] / (\gamma_3 + \gamma_4) \quad (4.5)$$

For the assessment of economic loss of the hospitals, only the flooding hazard was taken into consideration. Hospitals are irregular systems where the financial cost of the nonstructural elements may be (is often?) really high and consists of up to 90% of the whole facility value. For this reason, an economic loss assessment was carried out with the hypothesis solely considering the medical equipment aspect. As the first step, data extraction from the Florence Health System technology database was made by the Clinical Engineering department. Secondly, by using the form reported in Figure 4.3 and the database field "hospital area of use", it was possible to link specific medical equipment to precise locations within the hospital area, see Figure 4.4.

The final step examined the economic estimation of each technology component according to the estimated purchasing cost, see Figure 4.5. Once the correct linkage of device—cost—hospital location is obtained, the economic loss due to total cost of exposed devices to floods can be effectively estimated.

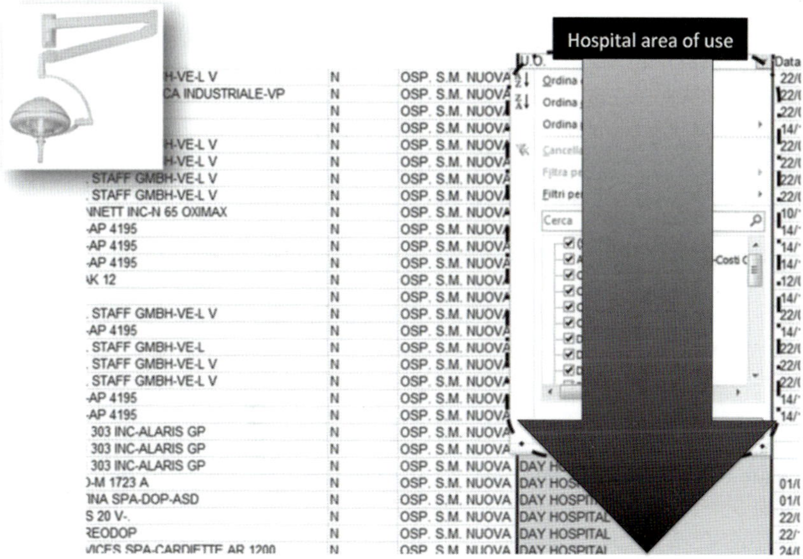

FIGURE 4.4 Example of medical equipment database in hospitals (Azienda Sanitaria Firenze "ASF" DB®).

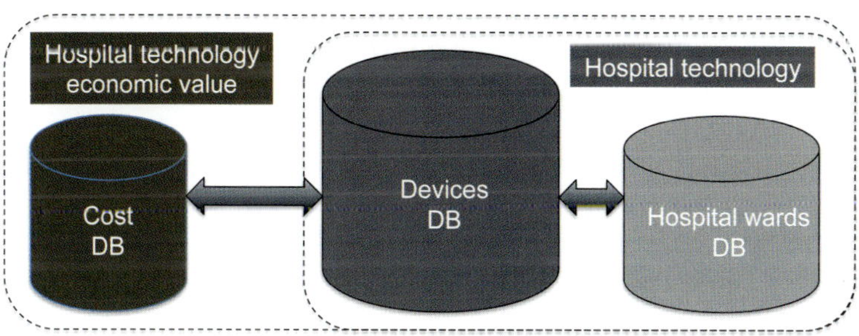

FIGURE 4.5 Economic assessment model based on a specific database implementation.

4.3 RESULTS

Five experts were interviewed for the definition of the matrix coefficients: two hospital engineers, one technical director, one physician, and one health services director. For the highest expertise positions, the number of experts interviewed might be representative: 50% for technical directors and 20% for health service directors. Regarding the lowest expertise positions (hospital engineers and physicians), the lower representativity is compensated with a lower specific weights. The questionnaire consisted of 256 semistructured questions which considered both the expertise and confidence levels of participants. The validation, carried out by considering experts and opinions' reliability, showed how all the experts were within the 0.05 confidence range and only three opinions out of 265 resulted to be slightly higher than 0.05. All the values were accepted as proper Leontief matrix coefficients. The highest *dependency index* "δ" is obtained for the surgical area while for the *gain index* "ρ", the highest value is for the power system.

4.3.1 Seismic Scenario

General consideration coming from the field analysis of townscapes revealed a number of situations of high seismic vulnerability related primarily to the following aspects: complex building forms, poor state of maintenance, probable loss of architectural details such as poorly secured façade moldings and perched bell-cotes. Blockage of access roads by the collapse of buildings that flank them, or as a result of mountain landslides, especially seismically induced slumping and debris flows.

In order to assess hospital performance, a magnitude 6.0 $\mathbf{M_W}$ event was hypothetically considered. The provincial health response degradation was about 33.39% for the daytime scenario while for the night/holiday it is 63.17%. The higher degradation for the night/holidays scenario is due to organizational reasons characterized with lower presence of staff, number of hospital, and ICU activated beds. In this case, most medical personnel are "on call" (MOVE handbook, 2011). By considering intrinsic security, the results showed that 30% of hospital inpatients would be adversely affected to the point of being unsafe (see Figure 4.6 Miniati and Iasio, 2012). Forty percent of the total area of

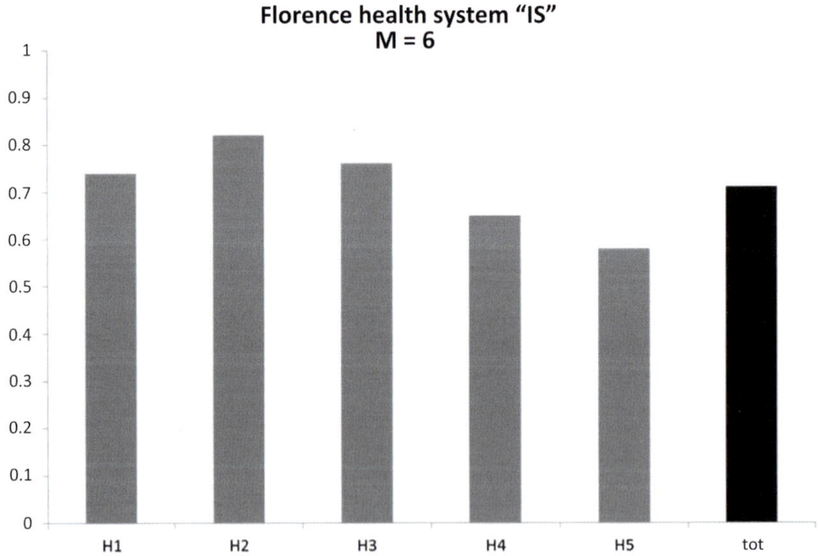

FIGURE 4.6 Intrinsic security degradation regarding the Florence hospital system (Miniati and Iasio, 2012).

Hospital H5, which is the closest hospital to the hypothetical seismic epicenter, is predicted to become insecure.

4.3.2 Flood Scenario

The following section considers the results according to the scenario provided by the Florence November 4th 1966 flood which caused 34 deaths. As defined for the seismic case, each hospital is named as H1 to H5 while the ambulance rescues headquarter is refereed with the letter "A". The definition of the failure vector was evaluated by applying the specific rapid vulnerability evaluation form reported in Figure 4.3 according to the 1966 floodwater levels map described in Figure 4.1.

This showed high exposure regarding the ambulance management headquarters and H3 which are in the 0–4 and 4–5 m depth floodwater area, respectively. Hospitals H2, H4, and H5 are located outside the map area, which is focused on the Florence downtown. These hospitals are not exposed to the 1966 flood scenario. H3 and "A" evaluation forms reveal 100% functionality degradation due to the loss of all the basic installations, including the server room for the ambulance management offices. For the municipal hospitals H1, H2, and H4, the general degradation is reported in Figure 4.7. Moderate vulnerability was estimated for hospital accessibility (caused by the loss of ambulance offices) and high vulnerability for plumbing systems, H5 however only suffered negative effects on ambulance accessibility.

Regarding the holistic performance assessment, the HTC for the province of Florence is reported in Figure 4.8. The total pre-event HTC passed from an initial value of 26 to a post-event level of 18 and 1.8 for day and night scenarios, respectively (daytime degradation of 30%, nighttime performance loss of 93%).

Figure 4.9 reports the vulnerability levels according to each hospital for day and night scenarios. Degradations of 20% for H1, H2, and H4; 10% for H5; and 100% for H3 which would completely lose its treatment capacity. For the holiday/night scenario estimated losses are as follows: 90% for H1, H2, H4, and H5 hospitals; 100% for H3.

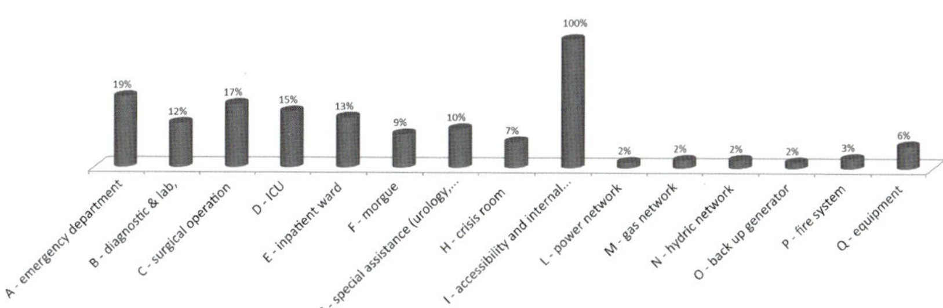

FIGURE 4.7 Inoperability output vector for Hospital H3.

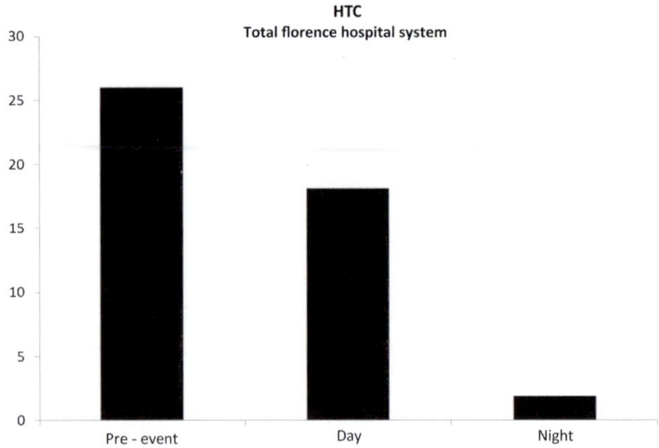

FIGURE 4.8 Estimated reduction of hospital treatment capacity (HTC) due to the 1966 Florence flooding scenario at the provincial level.

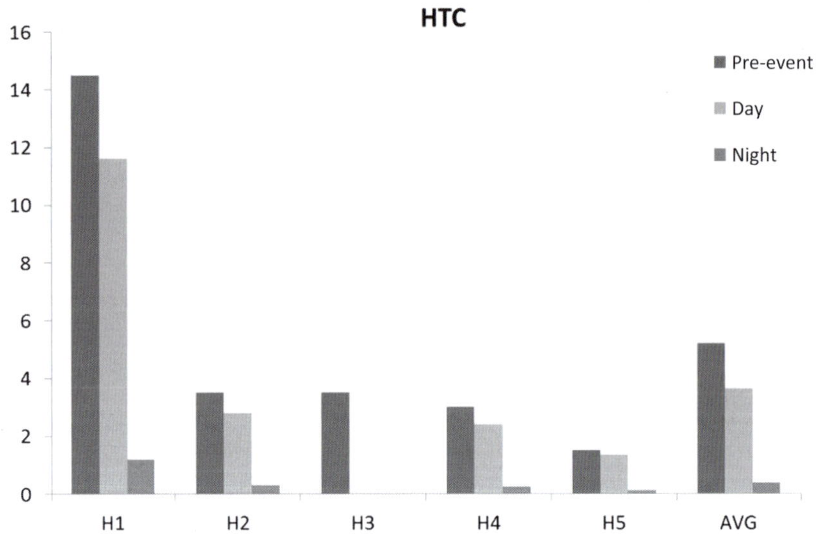

FIGURE 4.9 Estimated reduction of hospital treatment capacity (HTC) due to the 1966 Florence flooding scenario at the hospital level.

With regard to IS, the level of the whole system response shows that no significant differences were found between the day and night scenarios, both suffering a 30% degradation of patients safety at the ICU and inpatients areas. The estimated loss of devices at H3—the only structure with exposed medical equipment—consisted of almost half of all medical devices being lost during the flood scenario (47%). The most affected equipment belongs to the emergency department, diagnostic area, laboratory, and ICU.

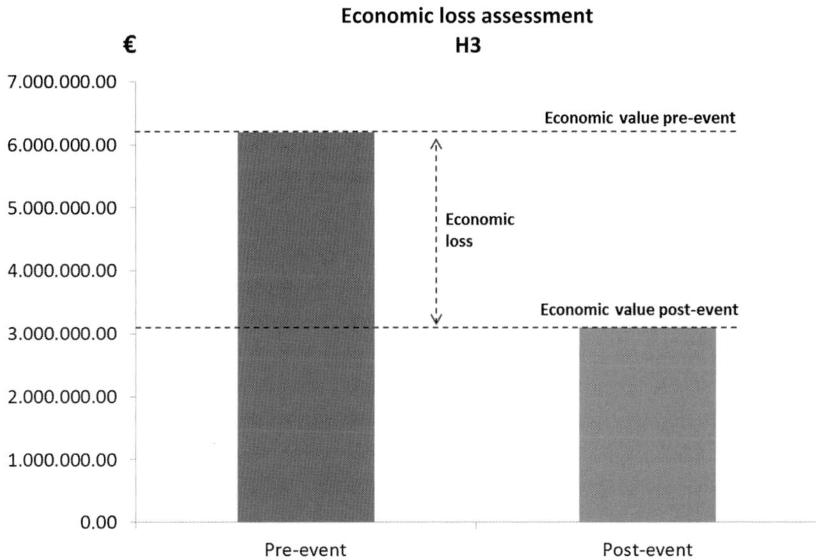

FIGURE 4.10 Estimated economic loss due to the 1966 Florence flooding scenario at hospital H3.

Figure 4.10 reports the subsequent economic loss derived by the technology toll which corresponds to the loss of half the total assets. This is due to the damage of the most expensive technology typical for imaging such as RMI, radiology, and CT equipment.

4.4 CONCLUSION

The rapid assessment procedure described in this paper provides a quick and easy methodology to assess the vulnerability to floods and earthquakes of health facilities by providing information on both the Florence area and each hospital. As reported in the MOVE project handbook, in the case of earthquakes and floods, the results showed that health system responses strongly depend on temporal as well as geographical exposure, as the staffing and organizational aspects strongly depend on the time and the day of occurrence of the event. Moreover, the use of HTC index with the development of the new index of IS allows the estimation of both the strategic and sheltering functions of hospital systems.

In case of floods, the most vulnerable infrastructure within the Florence hospital system is the patient admission component which causes serious knock-on problems for the emergency department in terms of patient's acceptance and discharging. The driving causes of this vulnerability are issues with the ambulance management service and the precarious and exposed location of vital infrastructure, such as the servers for the informatics system located at ground floor level. At the building level, H3 HTC is almost null in aftermath of the flooding event because of the loss of technology and basic lifelines systems and serious problems regard the intrinsic security for ICU beds and inpatients evacuation.

In the case of the seismic scenario, although the Florence treatment capacity seems appropriate, the intrinsic security suggests a moderate level of risk with a dangerous rate of 30% of the total. These conclusions have led us to believe that integrated planning focusing on improving both hospital system resilience (nonstructural and structural elements) and road viability, in addition to proper planning, might reduce the assessed vulnerability level.

Finally, further work includes testing the methodology to other modern hospitals, carrying out a scientific validation of the model in order to precisely define the correct numerical values and variables which have been, so far, currently, obtaining from scientific literature.

REFERENCES

Barton, D.C., Eidson, E.D., Schoenwald, D.A., Stamber, K.L., Reinert, R.K., 2000. Aspen-EE: an Agent- Based Model of Infrastructure Interdependency. Sandia Report SAND.

Bologna, S., Casalicchio, E., 2007. La Simulazione: uno strumento a supporto della protezione delle infrastrutture critiche. Safety & Security Journal (3).

Caballeros, O., México, Z.R., 1994. The Impact of Natural Disasters on Developing Economies: Implications for the International Development and Disaster Community. www.crid.or.cr/digitalizacion/pdf/eng/doc4907-a.pdf.

Commission of the European Communities, December 12, 2006. Proposal for a Directive of the Council on the Identification and Designation of European Critical Infrastructure and the Assessment of the Need to Improve Their Protection. 787. COM, Brussels.

De Boer, J., 1999. Order in chaos: modeling medical management in disasters. European Journal of Emergency Medicine 6 (2), 141–148.

Dubois, D., Prade, H., 1998. Possibility Theory: an Approach to Computerized Processing of Uncertainty. Plenum Publishing Corporation.

Federal Emergency Management Agency, Washington, D.C.: FEMA, Febrero 2003. Inundaciones, Antecedentes Y Preparación.

Haimes, Y.Y., Jiang, P., 2001. Leontief based model of risk in complex interconnected infrastructures. Journal of Infrastructure Systems, 1–12.

Iadanza, E., Dori, F., Miniati, R., Rossi, E., 2009. SEMIO: an expert system for monitoring critical hospital infrastructures. In: Proceedings from the World Congress in Medical Physics and Biomedical Engineering. 11th International Congress of the IUPESM. Sept 7 – 12, Munich Germany.

International Federation of Red Cross and Red Crescent Societies (IFRC). 2002. World disasters report 2002: Focus on reducing risk. Geneva: IFRC.

Kosko, B., 1994. The probability monopoly. IEEE Transactions on Fuzzy Systems 2 (1), 32–33.

Lupoi, G., Franchin, P., Lupoi, A., Pinto, P.E., Calvi, G.M., 2008. Probabilistic Seismic Assessment for Hospitals and Complex-Social Systems. IUSS Press, Pavia, 172 pp.

Miniati, R., Iasio, C., 2012. Methodology for rapid seismic risk assessment of health structures: case study of the hospital system in Florence, Italy. International Journal of Disaster Risk Reduction 2, 16–24.

Miniati, R., Iasio, C., Alexander, D., 2011. Vulnerability to Earthquakes and Floods of the Healthcare System in Florence, Italy. D.4.2 Handbook of Vulnerability Assessment in Europe. MOVE Project: Methods for the Improvement of Vulnerability Assessment in Europe. Seventh Framework Programme Cooperation Theme 6 – Environment (including Climate Change) Collaborative Project – GRANT AGREEMENT No. 211590.

Organización Panamericana de la Salud, Washington, D.C, 2006. Hospitales seguros ante inundaciones. Serie Mitigación de Desastres.

PAHO – WHO, 2000. Principles of Disaster Mitigation in Health Facilities. Disaster Mitigation Series. Washington DC.

Panzieri, S., Setola, R., Ulivi, G., 2005. An approach to model complex Interdependent infrastructures. In: Proceedings from the 16th IFAC World Congress 2005, Praha, July 4th–8th, 2005.

Salski, A., Fränzle, O., Kandzia, P., 1996. Introduction to the Special issue fuzzy logic in ecological modelling. Ecological Modelling 85, 1–2.

Sergio Castenetto, Mirella Sebastiano: Mugello, 29 giugno 1919: sui luoghi del terremoto. Regione Toscana, Firenze; 2004.

Setola, Roberto, 2007. Availability of healthcare services in a network-based scenario. International Journal of Networking and Visual Organisations 4 (2).

Setola, R., De Porcellinis, S., 2008. A methodology to estimate input-output inoperability model Parameters. In: Critical Information Infrastructures Security, Lecture Notes in Computer Science. Springer, Heidelberg.

World Health Organization, 2006. Health Facility Seismic Vulnerability Evaluation. A Handbook. World Health Organization, Europe Office, Copenhagen, Denmark.

Chapter 5

Vulnerability Assessment to Heat Waves, Floods, and Earthquakes Using the MOVE Framework: Test Case Cologne, Germany

Torsten Welle*, Yaella Depietri*,‡, Marjory Angignard†, Jörn Birkmann*, Fabrice Renaud* and Stefan Greiving†

*United Nations University, Institute for Environment and Human Security (UNU-EHS), Bonn, Germany, †Institute of Spatial Planning, Technical University of Dortmund, Dortmund, Germany, ‡Institut de Ciècia i Tecnologia Ambientals (ICTA), Universitat Autònoma de Barcelona (UAB), Barcelona, Spain

Chapter Outline

Assessment of Vulnerability to Natural Hazards. http://dx.doi.org/10.1016/B978-0-12-410528-7.00005-9

5.1 INTRODUCTION

Vulnerability assessment is an important first step toward effective risk reduction and climate change adaptation. Only if the population and decision makers know where and how vulnerable the system is and which social–economic, physical, and environmental factors play a major role in it, adequate measures can be implemented to reduce vulnerabilities to hazards and to adapt also to climate change. This study contributes to two different domains: primarily the scientific area, since it joins the research line of practical measurement approaches of vulnerability and disaster risk reduction (Maskrey, 1984; Chambers, 1989; Pelling, 1997; Cardona, 2001; Birkmann, 2006; IPCC, 2007); and secondly the policy area as this study provides information of the spatial distributions of vulnerability of the Cologne urban area to different natural hazards upon which the authorities need to take actions. The study is in this way in line with the objectives of the "Hyogo Framework for Action 2005–2015" which identifies risk assessment, monitoring and early warning as a current gap and challenge and which also calls for an increase of the "reliability and availability of appropriate disaster-related information to the public and disaster management agencies in all regions" (United Nations, 2005).

Vulnerability assessment needs to be based on a systematization and conceptualization of vulnerability describing the main linkages between the different components of risk (Downing, 2004). The MOVE framework (see Birkmann et al., 2013) was used as a conceptual framework in this study and its applicability was tested through assessing the vulnerability of the population of Cologne to the most relevant three hazards which affect the urban area. This paper presents the assessment of social vulnerability toward heat waves and floods on city district level, with some results for the environmental dimension of flood risk, as well as an approach to assess the institutional vulnerability of Cologne to earthquakes.

5.2 NATURAL HAZARDS WITHIN THE STUDY AREA

The Cologne urban area is situated in central-western Germany at 50°52′ North and 7°05′ East. It lies in the Federal State of North Rhine Westphalia (NRW) along the river Rhine (see Figure 5.1). The city is located along the Cologne

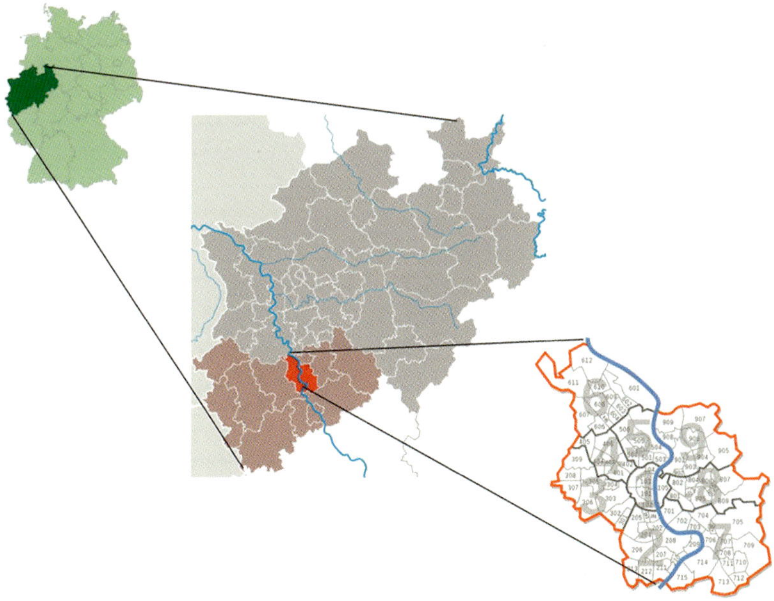

FIGURE 5.1 The federal state of NRW situated in Germany (upper left) and the state of NRW including the Cologne urban area (highlighted in red).

Bight which is an orographic depression between the central German uplands, in the southeast, and the German lowlands, in the north. The Cologne Bight forms a plain, geomorphic delta which reaches to the northwest and slopes toward the lower Rhine lowlands. It is surrounded by the *Bergisches Land* in the east and the *Siebengebirge* in the south. The study area has a temperate-oceanic climate with relatively mild winters and warm summers. The average annual temperature ranges between 9.5 and 10.5 °C, and the average annual precipitation ranges between 750 and 900 mm (Kuttler et al., 1997). Cologne is the largest city in NRW and within the Rhine-Ruhr Metropolitan Area with around one million inhabitants and the fourth largest city in Germany. The city is divided into 85 city districts which are further subdivided into 370 city quarters. From a socio-economic point of view, the vast majority of the lowest status neighborhoods are clustered into two parts: the larger one is located on east side along the river Rhine and just beyond, and the second one is located in the northwestern part of the city (Wolf, 2002). The districts with the highest social status are more dispersed but mainly located at the borders of the city (Wolf, 2002).

Heat waves, floods, and earthquakes are the main hazards that have affected the city in the past and can potentially affect it in the future. Heat waves can lead to major threats in densely populated cities, also in northern Europe, as happened in the 2003 extreme heat summer event, and might become more frequent due to observed and projected increase in the frequency and magnitude of warm daily temperatures (IPCC, 2012). Flood is the second hazard being

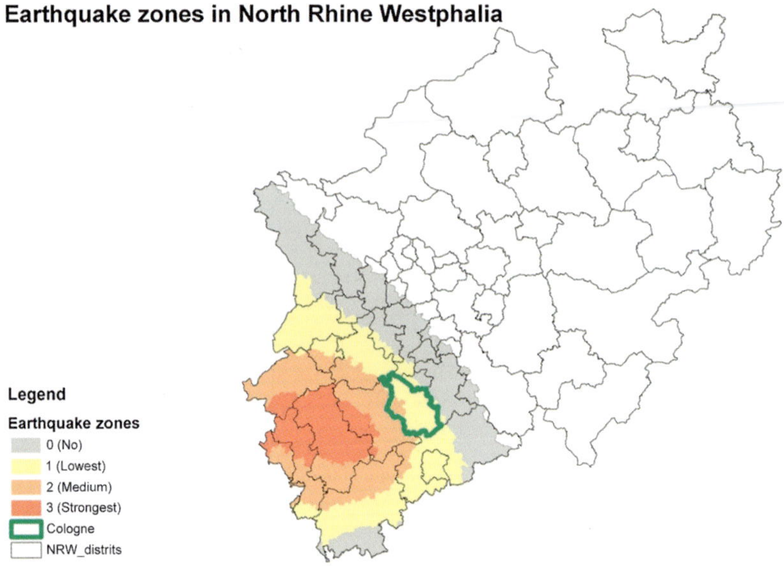

FIGURE 5.2 Different earthquake zones in NRW and highlighted the administrative unit of Cologne. The different earthquake zones represent the ground acceleration (a) and thus intensity intervals (I) of earthquakes according to the European macroseismic scale (EMS): Zone $1=a=0.4\,\text{m/s}^2=6.5\leq I<7$; Zone $2=a=0.6\,\text{m/s}^2=7\leq I<7.5$; Zone $3=a=0.8\,\text{m/s}^2=7.5\leq I$ (DIN 4149). *Own illustration based on data from the Geological Survey in NRW.*

considered as it has a long history of impacts in Cologne crossed by the largest river in Germany, the river Rhine. The city is also at risk of earthquakes. The Cologne Bight is in fact part of the Rhenish earthquake zone (see Figure 5.2) which ranges from Basel, in Switzerland, to the Benelux countries.

5.2.1 Heat Waves

Heat waves are extreme events in which hot temperature in summer months persist for a relatively long period of time. Impacts include human mortality and morbidity, costs for regional economies, and ecosystems degradation (Meehl and Tebaldi, 2004; Koppe et al., 2003). While there is no universal definition for heat waves, these are most commonly identified as events occurring when temperatures exceed both daytime high and night-time low thresholds (Vescovi et al., 2005) and this for a prolonged period of time. More specific definitions of heat waves require that thresholds are defined differently according to local conditions.

The Urban Heat Island (UHI) effect reflects the temperature difference between an urban area and its more rural surroundings and is mainly due to heat stored in buildings and in sealed roads (Gabriel and Endlicher, 2011). This effect, together with the higher concentration of air pollutants in cities (Wilhelmi and Hayden, 2010; Eliasson and Upmanis, 2000), aggravates the negative health related impacts of heat waves such as "skin eruptions, heat fatigue, heat cramps, heat syncope, heat exhaustion, and heat stroke" (Koppe et al., 2003). Causes of death or harm are linked to the age and fitness of elderly and to hazard factors such as the duration and intensity of the period of heat stress (Gabriel and Endlicher, 2011; Huynen et al., 2001).

According to Munich Re (2003), the European heat wave in 2003 caused over 20,000 deaths and an economic damage of 13,000M US$. Other sources estimate the death toll to have reaches 70,000 excess deaths of which most were elderly (Robine et al., 2008; Brücker, 2005; Vescovi et al., 2005). The international disaster database (EM-DAT) of the Centre for Research on the Epidemiology of Disasters (CRED) reported 9355 people killed and an economic damage of 1650M US$ due to extreme temperatures in August 2003 in Germany only (CRED-EMDAT, 2013). The Health Department of Cologne estimated that approximately 100–120 people were killed due to the extreme temperatures in August 2003. Climate projections predict that heat waves will become more intense, longer lasting, and more frequent (IPCC, 2012; Luber and McGeehin, 2008; Meehl and Tebaldi, 2004). It is therefore necessary to monitor and adequately plan to prevent negative effects on the population also at this higher latitude. The City of Cologne learnt from the 2003 event and, considering the potential reoccurrence of heat waves in the future, developed guidelines for the population on how to cope with heat waves. Additionally, city authorities together with the German Meteorological Service, the State office for nature, environment and consumer protection, and the urban drainage department of Cologne finalized a project to derive concepts and strategies for climate change adaptation.

5.2.2 Floods

Cologne is considered to be the most flood prone area in Europe. Flow's peaks at the Cologne gauge station documented an upward trend during the last century with an average increase of about 25% (Pinter et al., 2006). One of the main drivers of this change, which leads to the magnification of flooding in the area, seems to be mainly anthropogenic or due to climate change. The 10 years average (2001–2010) mean tide level of the Rhine in Cologne is 3.21 m, while first flood protection measures start at a level of 4.50 m, and at a level of 8.30 m ship ping traffic is interrupted (Urban drainage department City of Cologne, 2013). Table 5.1 shows all floods with more than 9.00 m level (Urban drainage department City of Cologne, 2013). Rhine floods in Cologne occur almost exclusively during winter months (Pinter et al., 2006) (see Table 5.1). Two floods which took place on the December 23rd, 1995 and on the January 21st, 1995 caused the highest damage in Cologne. The flood in 1993 caused four casualties and a damage of 65 M € whereas the economic losses in 1995 were about half of those in December 1993 (about 32.5. M €) as the population was better prepared, warnings were made in time and some measures were taken which largely reduced the damages and losses caused by the impacts of the flood (UNDINE, 2013).

TABLE 5.1 Floods Over 9 m Water Level in Cologne Since 1980

Date	Water Level
05.01.2003	9.71 m
25.03.2001	9.38 m
02.11.1998	9.49 m
21.01.1995	10.69 m
23.12.1993	10.63 m
29.03.1988	9.95 m
10.02.1984	9.11 m
30.05.1983	9.96 m
14.04.1993	9.81 m
08.02.1980	9.31 m

After the 1993 and 1995 floods, an agenda was developed in the beginning of 1996: "The flood protection concept for Cologne". These emphasized the importance of retention measures and aimed at strengthening them through the reconstruction of embankments in the hinterland, reshaping brooks in a natural way, unsealing areas of increasing seepage, and structural flood mitigations like mobile walls, dikes, or flood detention constructions. Until 2008, altogether 400 M € were invested in order to protect Cologne up to a flood of 11.30 m and the zones most at risk are protected from floods up to 11.90 m, reducing in this way exposure.

5.2.3 Earthquakes

The region of Cologne is crossed by the lower Rhine Graben. In this area, earthquakes are frequent: between 1976 and 2002 the Bensberg station measured more than 1300 small quakes in the area. However, most of them are not felt by humans and do not cause damages. In the recent history, only the 1992 Roermond quake reminded inhabitants of the Cologne area that they live under seismic threat. In this event, 1 person died and 25 got injured, 7200 buildings were damaged, and the total amount of losses amounted to 75 M € (Pelzing, 2008).

Seismologists reckon that quakes with a magnitude of 6.7 happened in the past, and could still hit the region. This would cause severe damages, as a study about seismic risk mapping for Germany revealed (Tyagunov et al., 2006). In this study, an assessment of potential seismic damage was carried out and estimated losses to about 790 M € for Cologne only. A large part of the building stock in Cologne was erected in the 1950s, after the destructions of World War II. Recent constructions had to comply with current building codes (among which are also paraseismic rules), but the older stock is considered not to be resistant to earthquakes. In addition, the presence of industrial sites near Cologne could lead to severe, secondary, and indirect effects.

Earthquakes cannot be predicted. However a dense network of stations (operated by different institutions such as the University of Cologne and the geological Survey of North Rhine Westphalia) is constantly monitoring the region. The map presented in Figure 5.2 shows the different earthquake zones in NRW (Grünthal et al., 2006).

5.3 VULNERABILITY ASSESSMENT WITHIN COLOGNE: METHODS

The City of Cologne thus is prone to different hazards as described above. Overall the focus of the authorities has been on the features of the hazard more than on those of the social-ecological system that contributes to its vulnerability. Regarding floods, the municipality runs a flood competence centre which is a large network of stakeholders bringing science and practice together for a holistic flood protection which include risk assessments. Additionally the Urban drainage department of Cologne is also dealing with flood protection measures which again aim at lowering mainly exposure. We suggest an alternative and complementary method to spatially calculate risk to floods of the Cologne population based on a wide set of social indicators accompanied by some results and reflections on the ecological dimension of risk, mainly relevant at river basin scale but that should also be considered by local authorities.

Heat waves are recognized as an important topic within the health department of the municipality but no concrete risk and vulnerability assessment based on city district are available to date. We present here a detailed social vulnerability assessment of Cologne at the city quarter level. For an integrated vulnerability assessment of social and ecological dimensions at the district level please refer to Depietri et al. (2013). The vulnerability assessments that we present in this study provide relevant information for risk reduction since the focus is not solely on the natural hazard sphere but also encompasses the social conditions and coping capacities of the system.

Vulnerability assessment to hazards focuses in general on the likelihood of injury, loss, and disruption of livelihood caused by an extreme event and/or by the obstacles in recovering from the disturbance that a system can potentially experience (Wisner et al. 2004, p. 13; Wisner 2002, p. 12/7). The focus of a vulnerability assessment lies on the identification of the variables that determine the vulnerability of people, represented by major differences in their (potential and revealed) susceptibility, exposure, and resilience. The analysis presented in this chapter is based on the MOVE Generic Framework, which addresses vulnerability and disaster risk to natural hazards from an integrative, holistic, and multidimensional point of view (see Figure 5.3). In this context, vulnerability refers to the propensity of exposed elements such as physical or capital

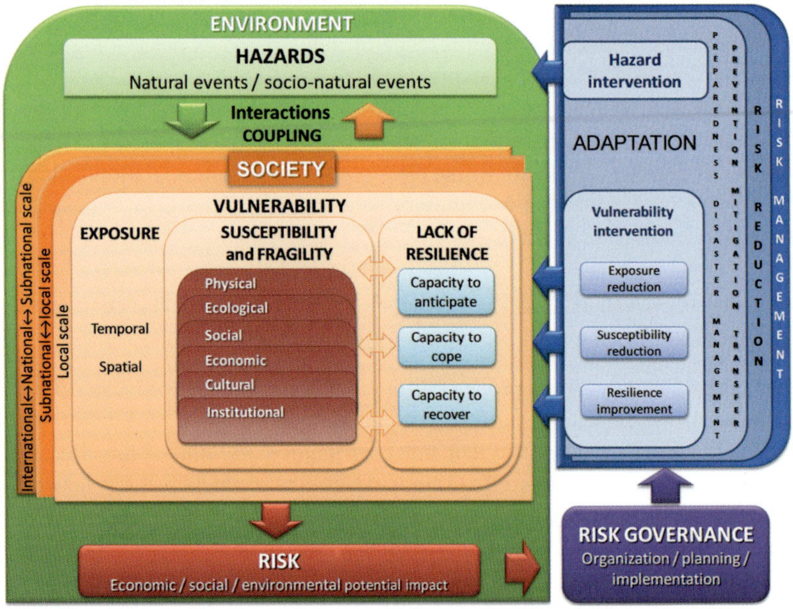

FIGURE 5.3 The MOVE Generic Framework used for vulnerability assessment (Birkmann et al., 2013).

assets, as well as human beings and their livelihoods, to experience harm and suffer damage and loss when impacted by single or compound hazard events and it is composed of key components such as exposure, susceptibility, and lack of resilience (or lack of societal response capacities) (Birkmann et al., 2013). Moreover, vulnerability is here described through different thematic dimensions such as physical, social, ecological, economic, cultural, and institutional.

The methods presented in this study assess different dimensions of vulnerability within Cologne related to three hazards. Precisely on the one hand: two quantitative assessment approaches focusing mainly on the social dimension of vulnerability toward floods and heat waves and on the other hand: one qualitative approach to the assessment of the environmental dimension of flood risk and one qualitative approach assessing the institutional dimension related to floods, heat waves, and earthquakes are presented. According to Birkmann et al. (2013), the social dimension of vulnerability is defined as the propensity for human well-being to be damaged by disruption of individual (mental and physical health) and collective (health, education services, etc.), social systems, and their characteristics (e.g., gender, marginalization of social groups). The environmental dimension refers to the potential for damage to all ecological and biophysical systems and their different functions and services (Birkmann et al., 2013). Finally the institutional dimension is defined as the potential for damage to governance systems, organizational form, and function as well as guiding formal/legal and informal/customary rules which might be forced to change due to the inadequacies they demonstrated while facing the hazard impacts.

The final assessment of vulnerability as described in the framework is translated into the following Eqn (5.1), both for the social dimension of vulnerability of the urban population to floods and heat waves:

$$V = E \times {}^{1}\!/_{2} \times (S + LoR) \tag{5.1}$$

where *V* is vulnerability, *E* exposure, *S* susceptibility, and *LoR* lack of resilience.

5.3.1 Assessment of Social Dimension of Vulnerability toward Heat Waves

The assessment of the social dimension of vulnerability toward heat waves comprises different datasets such as geo data and census data. Table 5.2 provides an overview of all data used, subdivided into the respective vulnerability key components.

TABLE 5.2 Data Used for each Factor of Vulnerability to assess the Social Dimension toward Heat Waves

Component of Vulnerability	Data Set	Data Source
Exposure	Thermal infrared satellite data of Cologne (June 30, 1993 at 9 p.m.)	Environmental department of Cologne
	Thermal infrared satellite data of Cologne (July 1, 1995 at 4 a.m.).	Environmental department of Cologne
	Number of inhabitants per city quarter	Statistical department of Cologne
	City districts (370 city quarters in Cologne)	Environmental department of Cologne
Susceptibility	Age groups (AG) 0–5 years + AG >65 years	Statistical department of Cologne
	Unemployment rate	Statistical department of Cologne
	Foreigners	Statistical department of Cologne
	Elderly households	Statistical department of Cologne
	City districts (370 city quarters in Cologne)	Environmental department of Cologne
Lack of resilience	Household members	Statistical department of Cologne
	City districts (370 city quarters in Cologne)	Environmental department of Cologne

The identification of adequate indicators and methods for each of the three factors of vulnerability is based on a literature review and on the results of expert workshops and stakeholder interviews. Several stakeholders and authorities of the City of Cologne were in fact contacted and interviewed in order to gather information and data, discuss the methodology and the outcomes of the study.

Following the MOVE approach exposure is defined as the number of people per administrative unit within the City of Cologne that can be potentially affected by extreme heat. The aggravating effect and differential exposure due to the UHI effect was integrated by using thermal remote sensing data which detect warm surfaces within the city. The thermal scans were conducted on June 30, 1993 at 9 p.m. for day temperatures and July 1, 1993 at 4 a.m. reflecting night temperatures. Exposure was then calculated using the number of inhabitants per city quarters multiplied by the normalized mean surface temperature per city quarter derived from thermal infrared satellite imagery. For the calculation of exposure in percentage the number of exposed population per city quarter was divided by the number of total population per city quarter.

Susceptibility indicators measure the predisposition of a society and ecosystems to suffer harm resulting from the levels of susceptibilities or fragilities of settlements and disadvantageous conditions and relative weaknesses related to ecological issues. For the case study of Cologne, the susceptibility indicators were extracted from census data. Based on literature review (e.g., Semenza et al. 1996; Brücker, 2005; Brown and Walker, 2008; Ebi, 2008; Fouillet et al., 2006; Klinenberg, 2001; Wilhelmi & Hayden, 2010), lessons learned from the 2003 heat event in Europe and stakeholders involvement age and health conditions, as well as socio-economic and socio-cultural factors are the main drivers that shape people's susceptibility. The following indicators for susceptibility were chosen for Cologne:

- Indicator 1 (Ind 1): age groups (AG) 0–5 years and AG >65 years (%)
- Indicator 2 (Ind 2): unemployment rate/city quarter (%), (Proxy: low income)
- Indicator 3 (Ind 3): foreigner/city quarter (%) (Proxy: problems in understanding of warning messages)
- Indicator 4 (Ind 4): elderly households/city quarter (%)

The indicators were weighted through expert judgments and aggregated based on Eqn (5.2):

$$S = (\text{Ind } 1 \times 0.6) + (\text{Ind } 2 \times 0.1) + (\text{Ind } 3 \times 0.1) + (\text{Ind } 4 \times 0.2) \tag{5.2}$$

Lack of resilience indicators measure the limitations in access to and mobilization of the resources of the social-ecological system, and the incapacity to respond by absorbing the impact. This factor of vulnerability includes the capacity to anticipate, cope, and recover in the short term. The selection of these types of indicators generally moves beyond quantitative aggregated demographic data and toward an understanding of knowledge, attitudes, and practices of individuals (Wilhelmi and Hayden, 2010). This type of information is not available in the form of statistics at the city quarter level and needs to be gathered through

Aditya AP N. 246

qualitative research. However, for this study, based on the understanding of safety nets in the form of social networks, the following assumption was made: the bigger the household (HH) is, the more HH-members are available, the more the elderly, invalids, and children living in the household can be assisted. This indicator is here used as a proxy for tight social networks measuring the capacity of the population to cope with extreme heat. For this indicator, data are available at the city quarter. Households with a single inhabitants of less than 30 years old were excluded in order to take into account only the older single households which are thought to be less able to cope with the hazard. The HH-sizes were then weighted based on experts judgment according to Eqn (5.2a) and (5.2b).

$$Coping\ Capacity = \begin{array}{l}(1*5HH+0,8*4HH+0,6*3HH\\ +0,4*2HH+0,2*1HH)/\ total\ number\ of\ HH\end{array} \tag{5.2a}$$

$$Lack\ of\ Resilience\ (LoR) = 1 - Coping\ Capacity \tag{5.2b}$$

To aggregate the factors exposure, susceptibility, and lack of resilience into vulnerability, Eqn (5.1) was used. Thereby all factors are expressed in the range between 0 and 1, where 1 represents the highest negative value. For example a susceptibility value of 1 would mean that every HH within a city district is susceptible.

5.3.2 Assessment of Social Dimension of Vulnerability toward Floods

Information about the datasets used for the assessment of the social dimension of vulnerability toward floods are provided in Table 5.3.

The exposure indicator for floods was derived by calculating the number of people who are affected by an extreme flooding event classified by an HQ 100. For this, spatial data at the administrative units of Cologne, demographic data per administrative units, and the flooding hazard layer, which is a product of a hydrological model showing the inundated areas and inundation depth, within the City of Cologne, were collected. The flooding hazard layer was combined with the population data in order to calculate the potential number of exposed people to an extreme flooding event (see Figure 5.10).

More in detail susceptibility was derived from census data and the selection of the indicators is based on the results of previous studies (Fekete, 2010). Steinführer and Kuhlicke (2007) provides a table of indicators that shows characteristics of high propensity to suffer harm from floods e.g., elderly might suffer more physical and health impacts and are less capable of following emergency measures promptly and effectively (Thieken et al., 2007). Very young people need more time to evacuate (Birkmann et al., 2008) and lower income groups have a lower degree of insurance available and are, in this way more, vulnerable than others (Steinführer and Kuhlicke, 2007; Thieken et al., 2007). Additionally, education and household size were identified as relevant indicators since people with lower levels of education support less and are less capable of performing emergency measures effectively (Steinführer and Kuhlicke, 2007; Thieken et al., 2007).

TABLE 5.3 Data Used for each Factor of Vulnerability to assess the Social Dimension toward Floods

Component of Vulnerability	Data Set	Data Source
Exposure	HQ 100 flooding hazard layer	Environmental department of Cologne
	Number of inhabitants per city quarter	Statistical department of Cologne
	City districts (370 city quarters in Cologne)	Environmental department of Cologne
Susceptibility	HH-type 1: includes households with children less than 6 years	Statistical department of Cologne
	HH-type 2: includes household-members solely between 6 and 59 years	Statistical department of Cologne
	HH-type 3: includes people older than 60 years by a household size of minimum two persons	Statistical department of Cologne
	HH-type 4: includes single households older than 60 years	Statistical department of Cologne
	City districts (370 city quarters in Cologne)	Environmental department of Cologne
Lack of resilience	Occupancy per household	Statistical department of Cologne
	City districts (370 city quarters in Cologne)	Environmental department of Cologne

Focusing on emergency measures, such as the evacuation ability of exposed people, the following assumption was made: regarding susceptibility, different households were characterized by people who might be in need of external help in case of a flooding event (named evacuation ability). This information is of high relevance for civil protection, since they are in charge of organizing the evacuation. The Statistical department of the City of Cologne provided data calculated by means of the HH Gen (Household generation method) software, which categorized all households into 11 classes based on their size and age of inhabitants. These 11 classes were again classified into four classes in order to guarantee an area-wide applicability: (see Table 5.4): HH-type 1 includes households with children less than 6 years; HH-type 2 includes household-members solely between 6 and 59 years; HH-type 3 includes people older than 60 years by a household size of minimum two persons; and HH-type 4 includes single

TABLE 5.4 Evacuation Ability of Various HH-Types in Cologne Based on a Household Survey

	Question Within the Household Survey: "Would You Be Able to Evacuate Without External Help?"	
HH-Types	**Yes**	**No**
HH-type 1	91.7%	8.3%
HH-type 2	95.3%	4.7%
HH-type 3	88.2%	11.8%
HH-type 4	60.5%	39.5%

households older than 60 years. As a second step, we carried out a household survey were we asked a question related to the individual evacuation ability of each HH-type. The result of this survey is shown in Table 5.4.

Both information, the number of HH-types and the results of the question regarding evacuation ability, were used to calculate the indicator percentage of HH who are not able to evacuate without external help. This indicator reflects the susceptibility of the individual city districts toward a flooding event (see Eqn (5.3)).

$$S_{floods} = (number\ oHH - type\ 1 \times 0.083) + (number\ oHH - type\ 2 \times 0.047) + (number\ oHH - type\ 3 \times 0.118) + (number\ oHH - type\ 5 \times 0.395) \quad (5.3)$$

Characteristics which strengthen the resilience of people are, according to Steinführer and Kuhlicke (2007), Birkmann et al. (2008), and Thieken et al. (2007), higher income and higher education as well as bigger household sizes or experience with the hazard such as floods. The assumption that the degree of coping capacity of several households is dependent on their individual experience with flooding events has also been documented (Steinführer and Kuhlicke, 2007). Inhabitants who already experienced a flood know how to react before, during, and after it and have awareness of its negative consequences such as physical or mental problems which follow an event. The indicator on flood experience allows identifying which city quarters the households show low capacity to cope with flooding and thus need more awareness training. The information needed for the development of the indicator "flood experience" is first derived on the occupancy of the individual households (variable x, see Eqn (5.4)) and the degree of exposure at the place of residence. Then more information was obtained through the households survey in which the following question was asked: "How many households have already experienced a flood at their place of residence?" By using the binary answers (Yes, No) to the mentioned question as a dependent variable and the occupancy as independent variable a logistic regression model was set up in order to assess the percentage of households per city district with flooding experience as a measure of resilience. Following logistic regression model Eqn (5.4) was used.

$$P(Y=1) = \frac{e^z}{1+e^z} \text{ with } z = b_0 + b_1 x_1 \tag{5.4}$$

The lack of resilience is calculated by using one minus flooding experience (see Figure 5.12). To aggregate all factors to vulnerability Eqn (5.1) was used.

5.3.3 Assessment of Environmental Dimension of Vulnerability toward Floods

Due to the regional scale at which regulating ecosystem services for flood mitigation in Cologne are produced and to the numerous hard infrastructures distributed along the river Rhine, the assessment of the direct link between the provision of these services and the benefits the urban population derives for flood mitigation was not a straightforward task. A quantitative assessment was not deemed appropriate to calculate ecosystem services given the high surface of the river basin and the high human intervention in the river. A qualitative approach was then chosen to assess the level of awareness of the ecological dimension of risk to floods.

The perception of local authorities about the environmental dimension of vulnerability was investigated through a questionnaires and the results discussed with respect to other related studies. Stakeholders' interviews were carried out between December 2011 and January 2012. Interviewees were identified at the city level amongst those institutions in charge or contributing to the planning and management of ecosystems and those responsible or active in the sector of human health also at the city level. A list of 25 relevant institutions and organizations was drafted. The list was discussed internally, benefiting from our previous experience of working with disaster risk in Cologne. Some of the institutions contacted via mail and/or phone refused to participate as they did not consider their daily work to be strictly relevant for our study, but, nevertheless, some of them pointed out other local authorities that they thought would be more appropriate to be involved in our analysis. We finally ended up with a list of seven institutions willing to be interviewed which covered exactly the set of dimensions we considered in our study (see Table 5.5).

The interviews were based on a questionnaire composed of open questions divided into two sections: the first set focused on the perceptions of the interviewees with respect to the role of regional/river basin ecosystems to mitigate the impacts of floods in Cologne; while the second focused on past, witnessed, and potential impacts of floods on the local urban and peri-urban ecosystem and its services (e.g., water supply, recreational activities) and thus, indirectly on the urban population. Each interview lasted from 40 to 60 min, was recorded and then transcribed. The transcribed text was analyzed making use of the Atlas.ti (ATLAS.ti Scientific Software Development GmbH, Berlin) software, which

TABLE 5.5 List of the Stakeholders Interviewed

Interviewee	Male (M)/ Female (F)	Position	Type of Institution	Sector
1	M	Project manager	Municipal	Environment
2	M	Head of department	Municipal	Landscape and urban green areas
3	M	Head of department	Municipal	Public health
4	F	Head of department	Municipal	Urban planning
5	M	Head of department	Municipal	Water management
6	M	Professor	University Institute	Public health
7	M	Project manager	Local branch of a national NGO	Forest management

supports qualitative data analysis. The software facilitates the cross comparison of texts by coding the information and allows for the generation of knowledge avoiding the reduction of the complexity contained in the data collected.

5.3.4 Vulnerability Assessment of the Institutional Dimension with respect to Heat Waves, Floods, and Earthquakes

The institutional dimension refers to the organizational form and function as well as guiding legal and cultural rules. This assessment approach differs clearly from those described above, since the performance and internal processes of individual authorities within a city were analyzed and no differentiation into exposure, susceptibility, and lack of resilience is feasible.

For the assessment of the institutional dimension, stakeholders from following authorities were involved: Environmental department, Statistical department, City development department, Parks and open space department, the Flood protection center, and the Geological survey of NRW were interviewed in order to receive an understanding of how processes and structures are perceived within institutions with respect to heat waves, floods, and earthquakes.

The indicators chosen were semiquantitative: the information is clearly qualitative but is assessed on a five level scale. Following 15 indicators were used which cover several aspects of institutional vulnerability: (1) principles,

(2) aim, (3) trust, (4) internal accountability, (5) external accountability, (6) justification, (7) representation, (8) access to information, (9) tolerance toward the process, (10) dialogue, (11) financial resources, (12) staff resources, (13) role of experts, (14) coordination, (15) cooperation. These indicators were derived from past projects, in which they had been tested and validated. Each indicator was applied to the three hazards considered in this case study (i.e., earthquakes, floods, and heat waves).

The indicators were not weighed but all represented on a graph (see Figure 5.12 for the results of institutional dimension).

5.4 RESULTS

5.4.1 Results for the Social Dimension of Vulnerability toward Heat Waves

For the spatial analysis and mapping, the values of the calculated indices were separated into five classes using the quantile classification method, which is a tool of ArcGIS 9.3.

Figures 5.4–5.7 show the results for exposure, susceptibility, lack of resilience, and vulnerability of social dimension toward heat waves at city quarter level. Thus, Figure 5.4 shows the result for the exposure which was dependent on the surface properties and their heat retention capacity for each city quarter in combination with the respective population. Each city quarters could be identified with the highest and lowest number of exposed people. In general, a heterogeneous pattern of exposure was measured where the city quarter east of the river Rhine show a lower degree of exposed people than the quarters on the west side.

The result for the susceptibility to heat waves in the city of Cologne is presented in Figure 5.5. The four indicators used for its measurement are based on socio-economic information obtained from the Statistical department of the City of Cologne. It is shown that the city centre of Cologne and the two city quarters Villen Viertel and Andreas Viertel are situated in the lowest susceptibility class since single and younger households, such as of students, inhabit these quarters. It is worth noting that the more susceptible city quarters are located in the outskirts of Cologne.

The lack of resilience, or the lack of coping capacity, is shown in Figure 5.6. Since this indicator is focusing on household size, an inverse picture compared to the susceptibility map is noticeable. It can be seen that the city centre has a high lack of coping capacity since mainly single households are located in it whereas higher coping capacities can be found in the outskirts of Cologne.

According to Eqn (5.1) all components of vulnerability are aggregated in one vulnerability index. Figure 5.7 illustrates the different vulnerability patterns regarding heat waves in Cologne. The most vulnerable city quarters are located

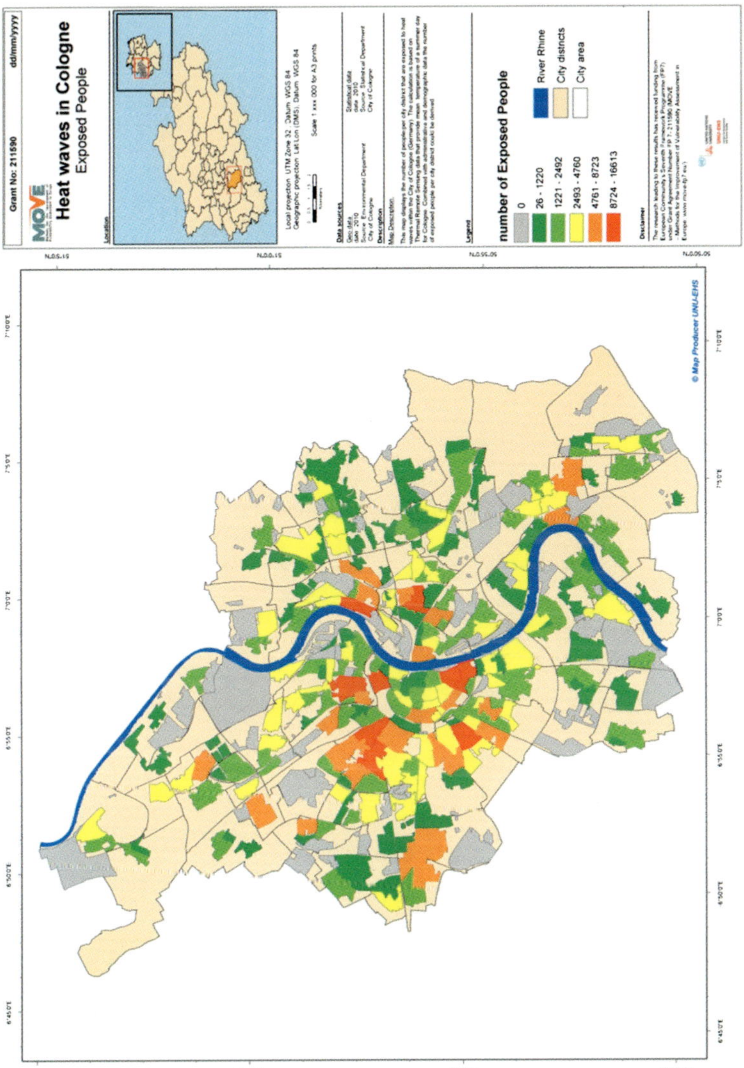

FIGURE 5.4 Exposure map of Cologne related to heat waves.

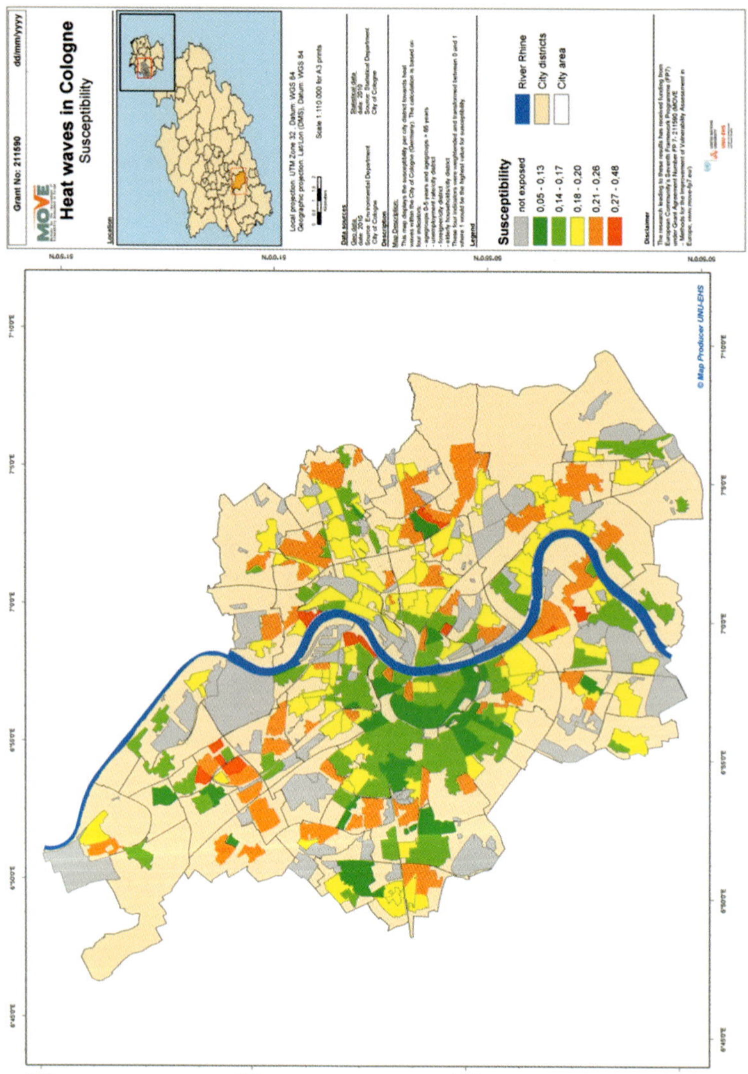

FIGURE 5.5 Susceptibility map of Cologne related to heat waves.

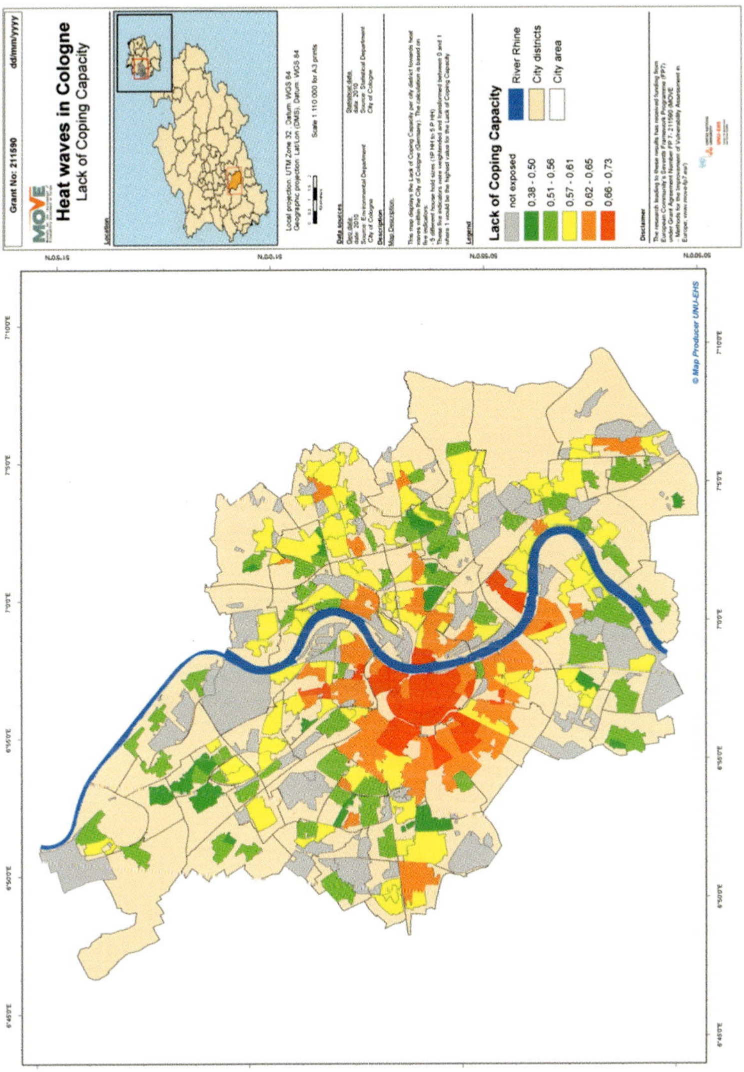

FIGURE 5.6 Lack of coping capacity map of Cologne related to heat waves.

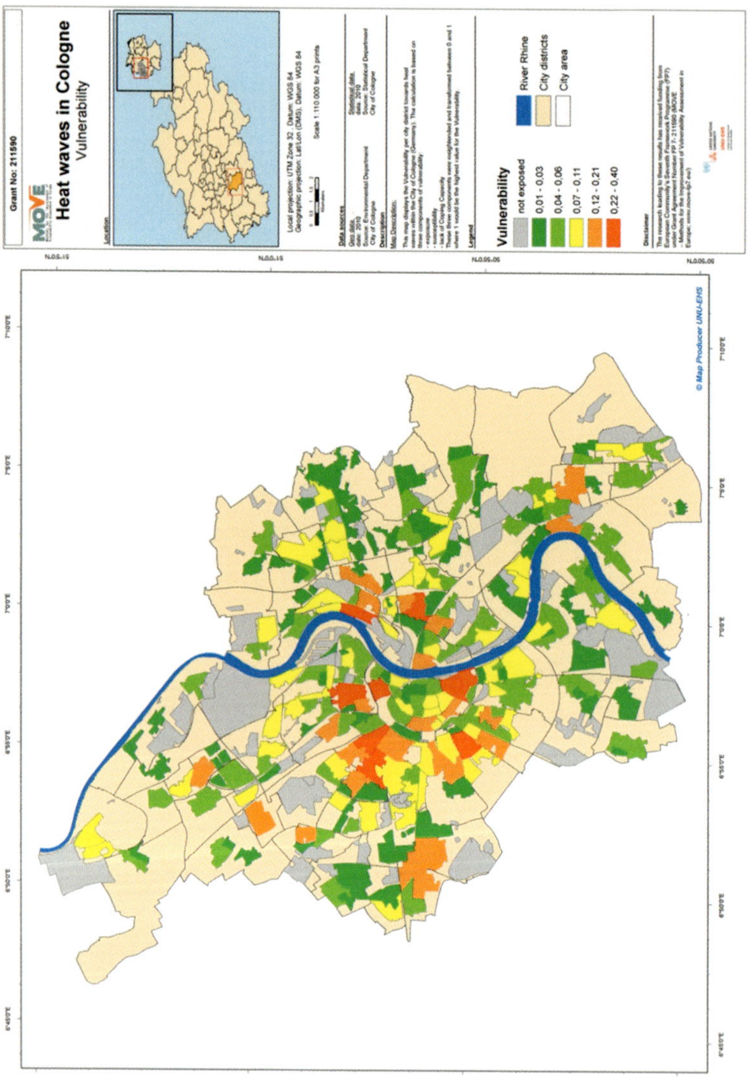

FIGURE 5.7 Vulnerability map of Cologne related to heat waves.

close to the city centre whereas the outskirts are in general less vulnerable. The western part of Cologne shows more vulnerable city quarters than the east of the Rhine.

5.4.2 Results for the Social Dimension of Vulnerability toward Floods

Figures 5.8–5.11 show the results for exposure, susceptibility, and lack of resilience and vulnerability of the social dimension of vulnerability toward floods.

Figure 5.8 shows the results for the flood (HQ-100) exposure in Cologne. Compared to the exposure to heat waves the calculation or detection of exposed areas and exposed people is more explicit since the hazard and its extent could be clearly identified based on the modeled hazard layer. Thus the most exposed people are within the city quarters that are located along the Rhine.

The susceptibility was calculated only for the exposed city quarters. The most susceptible city quarters result to be located east of the Rhine as well as in the North-Western part of Cologne. Less susceptible quarters could be identified around the city centre (Figure 5.9).

Figure 5.10 shows the results for the calculation of the lack of coping capacity toward floods in Cologne or inexperience with floods. The households with high experience are located in the south of Cologne (Rodenkrichen) and in the north of Cologne (Mülheim).

The overall vulnerability map with respect to floods is displayed in Figure 5.11. The hotspots of vulnerability are located along the river and in the northern part of the City. In the western part of Cologne low vulnerability values could be identified since these city quarters are less exposed, less susceptible and offer a higher degree of coping capacities as compared to those quarters located along the river Rhine.

5.4.3 Results for the Environmental Dimension of Vulnerability toward Floods

The results of the stakeholders' interviews regarding the ecological dimension of floods in Cologne are summarized in Table 5.6. Most of the respondents strongly agreed that the forest cover, the state of soils, and wetlands at the river basin scale have a positive effect on the reduction of the magnitude of floods in Cologne. Current upstream urbanization has therefore also an impact and increased flood risk between 1960 and 2010. It was mentioned that as agriculture rather decreased in the past decades, it is urban expansion the main driver of deforestation in the basin. To mitigate floods in Cologne the main strategy so far has been the creation of new retention areas, some of which are nearly to be constructed in Langel/Porz and Worringen. A retention area in Cologne is also constructed in Dungel, with 5 M cubic meters of volume for the retention of flooding waters. A project is just starting with a construction of a

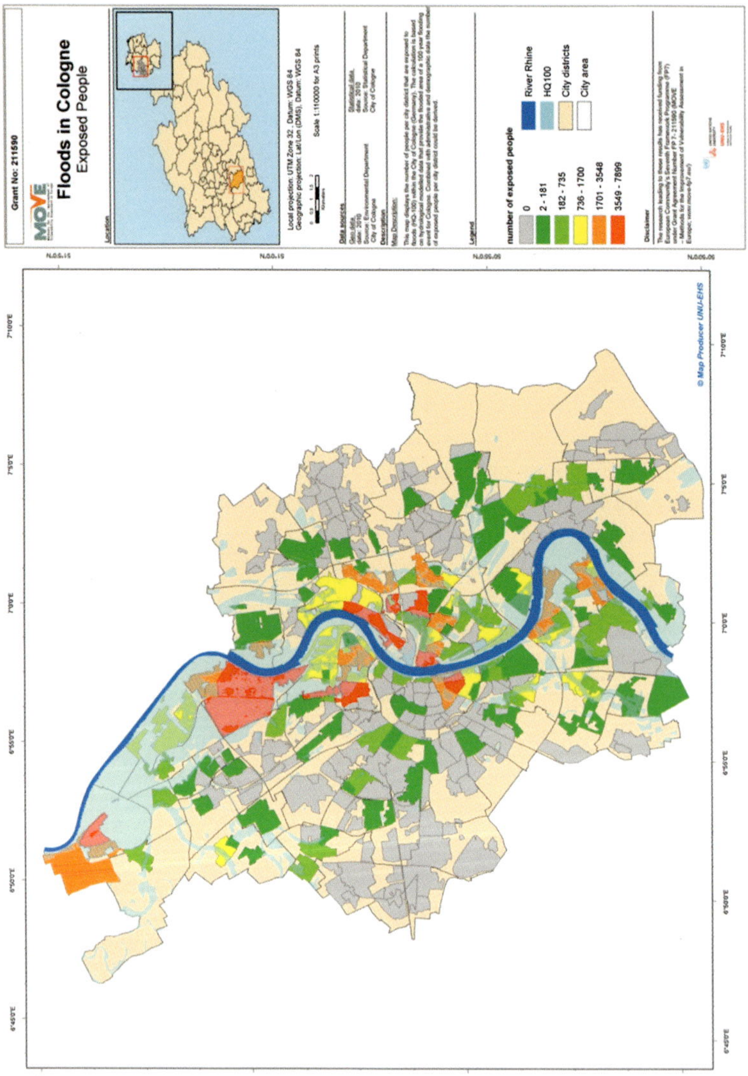

FIGURE 5.8 Exposure map of Cologne related to floods.

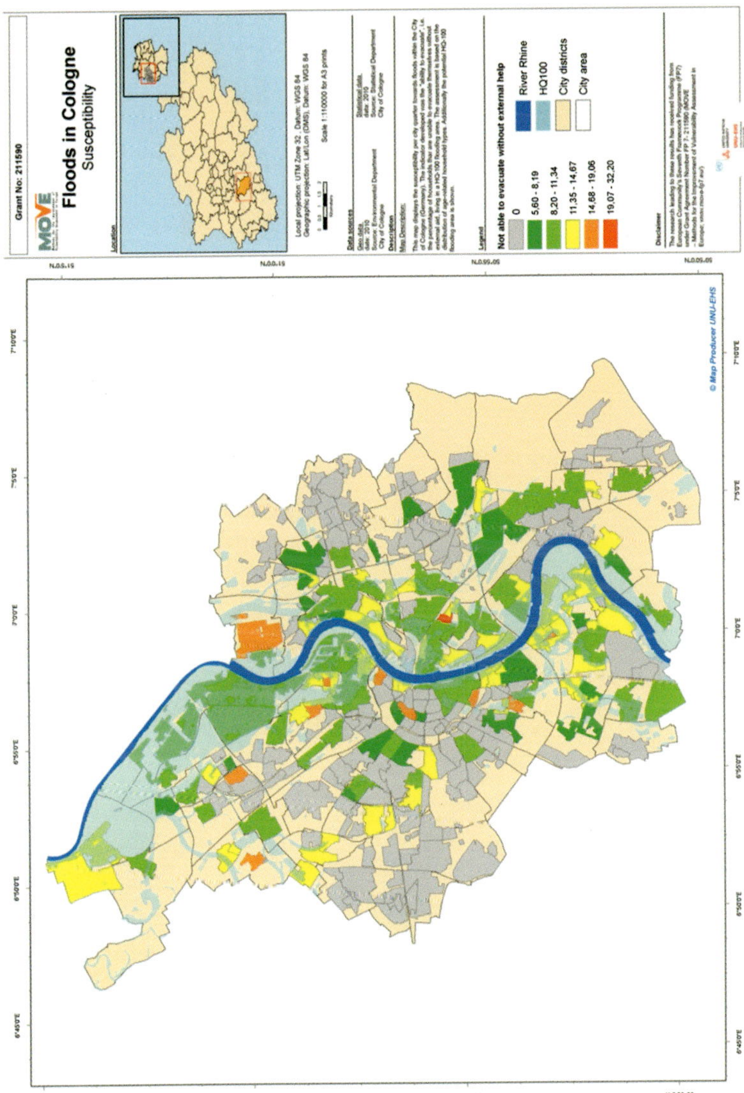

FIGURE 5.9 Susceptibility map for Cologne related to floods.

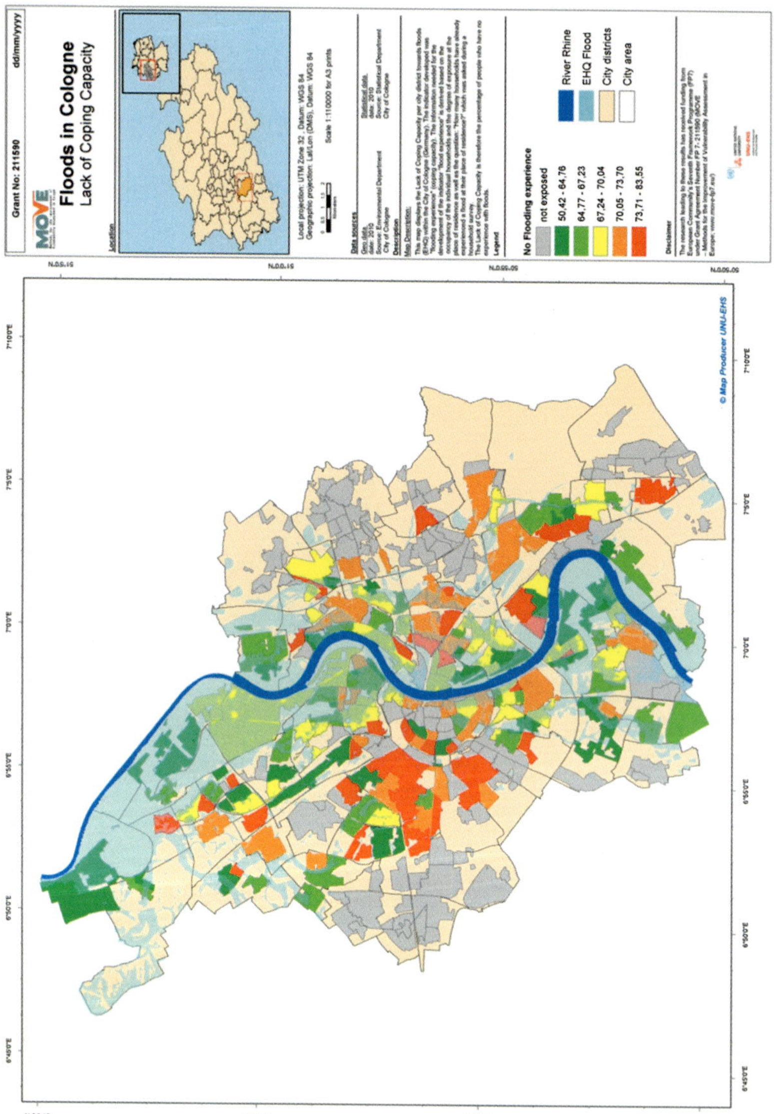

FIGURE 5.10 Lack of coping capacity map of Cologne to floods.

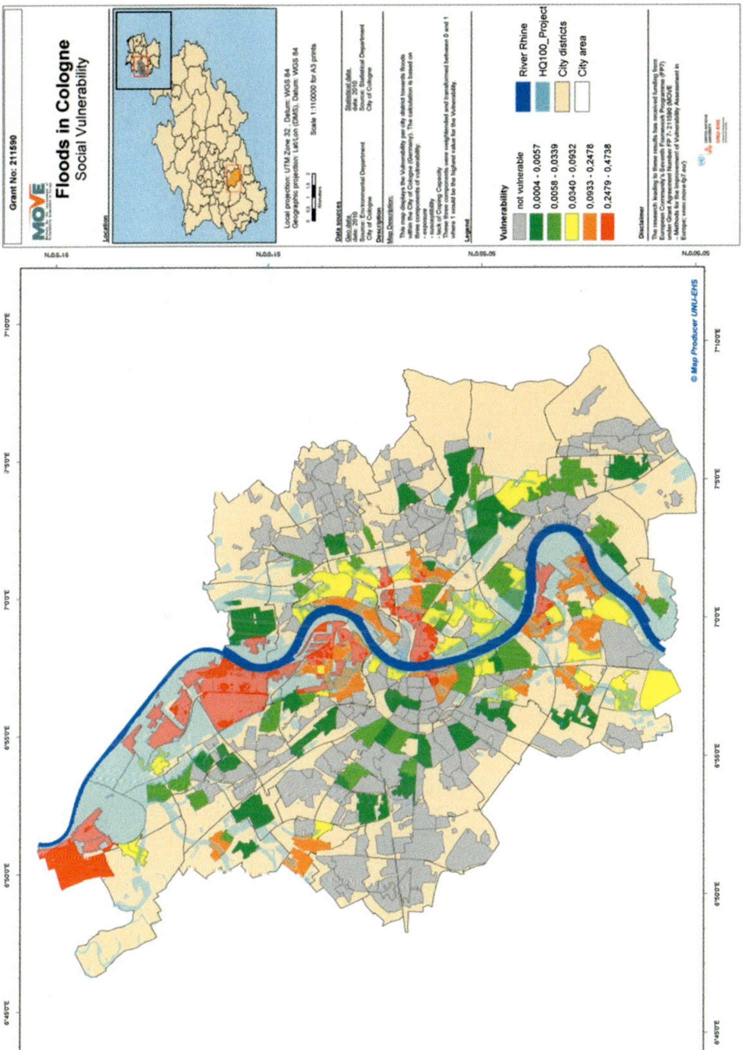

FIGURE 5.11 Vulnerability map of Cologne related to floods.

TABLE 5.6 Stakeholders' Perceptions on the Ecological Dimension of Floods in Cologne

Ecosystem	Ecosystem Service	Relevance (H = High, M = Medium, L = Low)	Benefit (River Basin Scale)	Impact (Local, Urban Scale)	Actions to Be Taken
Forests	Flood regulation	H	Moderate flood peak through better water infiltration, diminished runoff	Forested areas which are not adapted to persistent flooding, in case of an important flood, that take place each 100 or 200 years, can be damaged. Oaks would perish while alder and willow tree can resist to prolonged submergence. For long periods also poplar and street trees along the Rhine can be affected (especially for 5–6 weeks of submergence). However most of the vegetation does not get affected. Local soil sealing does not have an impact	Regulate urban expansion which is considered as the main driver of forest loss
	Recreation (city scale)			Recreational areas that might be flooded are Poller and Wiesen	
Soils	Flood regulation	H (at the river basin scale)	Water infiltration	Erosion and sedimentation	Plan to contain or reduce soil sealing
	Flash floods regulation	H (at the city scale)	Water infiltration The capacity of soils to retain water around Cologne is thought to be very high	Erosion and sedimentation in Cologne can happen but at insignificant rates and affects groundfloors and garages	Plan to contain or reduce soil sealing The type of soil should be used as an indicator of the retention capacity of the area rather than the type of land cover

Wetlands	Flood regulation	H	Water retention and release at slower rates	Wetlands and small water bodies can be affected in case of floods at the city scale	Preserve and restore wetlands upstream. The design of retention ponds should take into account the potential role of wetlands upstream. River embankments that disconnect the river from wetlands and flood plain should be avoided. See the International Commission for the Protection of the Rhine (ICPR) plan.
Freshwater	Water supply	L	Water supply	None	Pumps are immediately shut down in case of a flooding Monitor microbiological parameters
	Recreation	M	Recreational activities (e.g., fishing or swimming/bathing)	Water quality may deteriorate	Monitor microbiological parameters

Continued

TABLE 5.6 Stakeholders' Perceptions on the Ecological Dimension of Floods in Cologne—cont'd

Ecosystem	Ecosystem Service	Relevance (H = High, M = Medium, L = Low)	Benefit (River Basin Scale)	Impact (Local, Urban Scale)	Actions to Be Taken
Urban	Shelter and recreational activities	M	Shelter and recreational activities	Mud and fungus can grow causing unhealthy indoor environments. Industrial sites along the river are mostly safe with respect to flood hazards. Of higher risk are households and industrial sites where the water table can dangerously rise Some recreational areas can get affected but this does not cause important damages because of the wide variety of recreational areas in Cologne	Early warnings
Agricultural land	Food	L	Food provision	In the south of Cologne some agricultural land might be flooded	(N.B. Most of the agricultural land is not situated close to the river)

volume of 30M cubic meters also upstream of Cologne. Most of the projects mentioned regard the construction of infrastructures to contain flood nearby the city. Finally, according to the respondents, floods at the city scale might gain in importance in the future. For this type of risk, the rate of soil sealing in Cologne becomes an important factor to consider.

The second part of the questionnaire focused on the impacts of floods on the urban and peri-urban ecosystem and their services (see also Table 5.6). Most of the respondents agreed that water supply sector is not affected by flooding and that there is sufficient groundwater which is hardly affected by the hazardous event. Artificial lakes around Cologne on the other hand might be affected or contaminated. The industrial areas around the city are considered to be very safe with respect to floods, and agricultural land is not situated close to the river. It is thought by the majority of the respondents that there is not a major risk of pollution (anthropogenic risk) in Cologne related to floods. Some respondents nevertheless underlined their concern for oil tanks around Cologne which might be damaged during an event. Sedimentation around Cologne is not considered as a major problem. Regarding impacts to the vegetation, only rare floods occurring every 100 or 200 years can have an impact on the vegetation and recreational activities when the forest can be affected being submerged for long periods.

5.4.4 Results for the Institutional Dimension of Vulnerability toward Earthquakes, Floods, and Heat Waves

The answers of the stakeholders are represented in Figure 5.12.

The performance of risk governance processes is hazard dependent. It can be clearly seen that floods are already well addressed. The performances of the associated risk governance processes are seen to be good in most domains. Earthquakes, although authorities and stakeholders are aware of the risks, do not attract as much attention. If scientists and experts have a clear role, the lack of risk management policy appears as a major problem. Heat waves have a specific status in Cologne: attentions arouse only recently (after summer of 2003) and the willingness to address this issue was expressed immediately after. However, the risk management policies are still being defined and it is too soon to give an opinion on their performance. Uncertainties still have to be waived, and regular repetition of a similar study would be of interest to monitor the evolution.

5.5 DISCUSSION AND CONCLUSION

Different methods have been presented and applied to assess different dimensions of vulnerability of Cologne to the main three hazards that would affect the city. The results of vulnerability assessment for heat waves (Figure 5.7) and for floods (Figure 5.11), based both on Eqn (5.1), show that the exposure factor has a big influence on the overall vulnerability of the city. This is due to the fact that we multiply exposure with the weighted sum of susceptibility and lack of

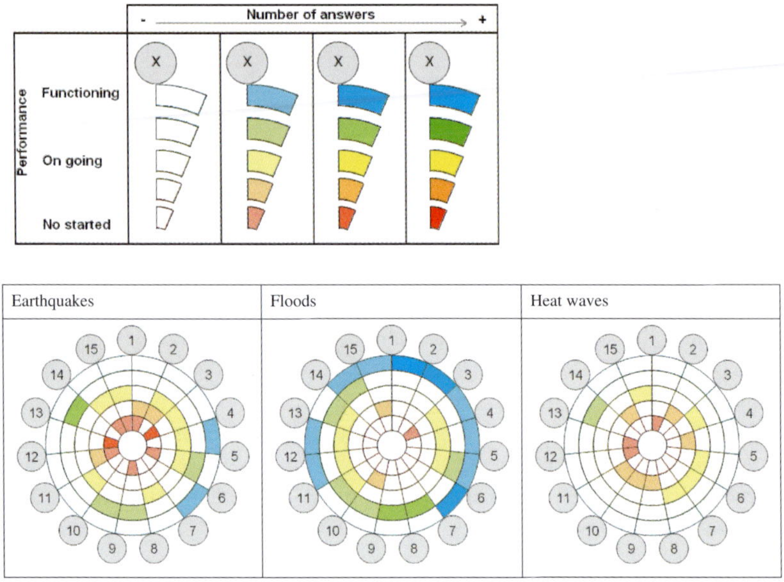

1.	Principles	6.	Justification	11.	Financial resources
2.	Aim	7.	Representation	12.	Staff resources
3.	Trust	8.	Access to information	13.	Role of experts
4.	Internal accountability	9.	Tolerance towards the process	14.	Coordination
5.	External accountability	10.	Dialogue	15.	Cooperation

FIGURE 5.12 Scheme of questions and results of the stakeholder discussion regarding institutional vulnerability.

resilience. Different weights for each factor of vulnerability could also be used based for instance on expert judgments or stakeholder opinions. The calculation of each of the different components of vulnerability however provided additional information on the reasons behind the distribution of vulnerability. Especially for heat waves for which susceptible groups appear to be more sparsely located around the city centre in areas which appear less exposed due partly to the minor UHI effect.

Indicators and indices, on the other hand, cannot fully meet the requirements of a theoretical concept. It has to be considered that the development of indices is a systematic and creative process in which representative data and variables is identified, gathered, obtained, or calculated and evaluated to solve certain problems. The advantage of an aggregated vulnerability index is the possibility of reducing a complex situation to one single value, which allows grasping the current situation at one glance. The individual factors and the aggregated vulnerability index which could be used for communication and can serve as first basis for decision-making. The validity of the indicators has been tested by external stakeholders.

Complementary information to the quantitative assessment of the social component of vulnerability to floods was derived through the perceptions of the relevant city level authorities on the environmental dimension. This moved the focus toward the more regional, river basin scale at which most of the responded identified important sources of provision of regulating services to mitigate flood risk in the city. Improving upstream ecosystem condition is then perceived as providing an effective strategy to reduce exposure to floods in Cologne. However, according to the literature wetland reclamation along the river might have some impacts on flood risk while land-use changes play a minor effect on large floods in large catchment areas such as the river Rhine (Disse and Engel, 2001; Hooijer et al., 2004; Pinter et al., 2006). A city regional approach as suggested in Depietri et al. (2014) might need to be strengthened in Cologne. Furthermore, according to the respondents, urban and peri-urban ecosystems and their services, on the other hand, would not suffer significant degradation or failure due to ordinary floods, lower than 100 HQ. Floods that affect Cologne take in fact place mainly in the winter time and when vegetation is dormant. Rhine summer floods on the other hand can affect vegetation in the basin (Vervuren et al., 2003) affecting recreational activities. As mentioned earlier, the assessment ecological dimension of vulnerability to heat waves was presented in Depietri et al. (2013). It resulted, at the contrary, that urban and peri-urban ecosystems play a central role in shaping the vulnerability of Cologne to heat waves and that improvements in their management could significantly lower the exposure as well as strengthen the resilience of the urban population.

Overall, it was shown that the MOVE theoretical framework can be applied to different hazards and to different dimensions (social, ecological, and institutional) and lead to policy relevant results. The presented methodological approach required data and resources which are currently used for different purposes by stakeholders in Cologne. The results are therefore also easily understandable and transferrable in the local decision making context. Stakeholder workshops, bilateral cooperation, and the results of the institutional assessment of vulnerability clearly showed that the perception and the relative importance of the three studied hazards are rather unbalanced for Cologne. In particular, the awareness of the city's population and local authorities with regard to heat waves and earthquakes needs to be strengthened.

REFERENCES

Birkmann, J., 2006. Indicators and criteria for measuring vulnerability: theoretical bases and requirements. In: Birkmann, J. (Ed.), Measuring Vulnerability to Natural Hazards—Towards Disaster Resilient Societies. United Nations University Press, Tokyo, pp. 55–77.

Birkmann, J., Krings, S., Renaud, F., 2008. Assessment of vulnerability to floods at local level with a special focus on human-environmental systems and critical infrastructures. In: Simonovic, S.P., Bourget, P.G., Blanchard, S.F. (Eds.), Presented at the 4th International Symposium on Flood Defence. Institute for Catastrophic Loss Reduction, Toronto.

Birkmann, J., Cardona, O.D., Carreño, L., Barbat, A., Pelling, M., Schneiderbauer, S., Kienberger, S., Alexander, D., Zeil, P., Welle, T., 2013. Framing vulnerability, risk and societal responses:the MOVE framework. Natural Hazards 67, 193–211.

Brown, S., Walker, G., 2008. Understanding heat wave vulnerability in nursing and residential homes. Building Research and Information 36, 363–372.

Brücker, G., 2005. The 2003 European heat waves. Eurosurveillance 10.

Cardona, O.D., 2001. Estimación Holística del Riesgo Sísmico utilizando Sistemas Dinámicos Complejos (Doctoral dissertation). Technical University of Catalonia, Department of Terrain Engineering. Available from: http://www.desenredando.org/public/varios/2001/ehrisusd/index.html.

Chambers, R., 1989. Vulnerability, coping and policy. Inst Dev Stud Bull 20 (2), 1–7.

CRED-EMDAT, 2013. The Centre for research on the epidemiology of disasters international Disaster database. Universite Catholique de Louvain, Brussels, Belgium. Available from www.emdat.be.

Depietri, Y., Welle, T., Renaud, F.G., 2013. Social vulnerability assessment of the Cologne urban area (Germany) to heat waves: links to ecosystem services. International Journal of Disaster Risk Reduction 6, 98–117.

Depietri, Y., Guadagno, L., Breil, M., 2014. Urban watershed services for improved ecosystem management and risk reduction, assessment methods and policy instruments: state of the art. SSRN Electronic Journal. FEEM Working Paper No. 101.2013.

DIN 4149: 2005–04: Bauten in deutschen Erdbebengebieten. Lastannahmen, Bemessung und Ausführung üblicher Hochbauten. Normenausschuss im Bauwesen (NABau) im DIN – April 2005, Berlin.

Disse, M., Engel, H., 2001. Flood events in the Rhine basin: genesis, influences and mitigation. Natural Hazards 23, 271–290.

Downing, T., 2004. What have we learned regarding a vulnerability science? In: Science in Support of Adaptation to Climate Change. Recommendations for an Adaptation Science Agenda and a Collection of Papers Presented at a Side Event of the 10th Session of the Conference of the Parties to the United Nations Framework Convention on Climate Change, Buenos Aires, 7 December 2004, pp. 18–21. Available from: www.aiaccproject.org/whats_new/Science_and_Adaptation.pdf.

Ebi, K.L., 2008. Healthy people 2100: modeling population health impacts of climate change. Climatic Change 88, 5–19.

Eliasson, I., Upmanis, H., 2000. Nocturnal airflow from urban parks-implications for city ventilation. Theoretical and Applied Climatology 66, 95–107.

Fekete, A., 2010. Assessment of Social Vulnerability to River-Floods in Germany.

Fouillet, A., Rey, G., Laurent, F., Pavillon, G., Bellec, S., Guihenneuc-Jouyaux, C., Clavel, J., Jougla, E., Hémon, D., 2006. Excess mortality related to the August 2003 heat wave in France. International Archives of Occupational and Environmental Health 80, 16–24.

Gabriel, K.M.A., Endlicher, W.R., 2011. Urban and rural mortality rates during heat waves in Berlin and Brandenburg, Germany. Environmental Pollution 159, 2044–2050.

Grünthal, G., Thieken, A.H., Schwarz, J., Radtke, K.S., Smolka, A., Merz, B., 2006. Comparative risk assessments for the city of Cologne – storms, floods, earthquakes. Natural Hazards 38, 21–44.

Hooijer, A., Klijn, F., Pedroli, G.B.M., Van Os, A.G., 2004. Towards sustainable flood risk management in the Rhine and Meuse river basins: synopsis of the findings of IRMA-SPONGE. River Research and Applications 20, 343–357.

Huynen, M.M.T.E., Martens, P., Schram, D., Weijenberg, M.P., Kunst, A.E., May 2001. The impact of heat waves and cold spells on mortality rates in the Dutch population. In: In: Martens, P., Schram, D., Weijenberg, M.P., Kunst, A.E. (Eds.), Environmental Health Perspectives, vol. 109, no. 5. , pp. 463–470. Environmental Health Perspectives 108, 463–470.

IPCC, 2007. Assessing key vulnerabilities and the risk from climate change. In: Parry, M.L., Canziani, O.F., Palutikof, J.P., van der Linden, P.J., Hanson, C.E. (Eds.), Climate Change 2007: Impacts, Adaptation and Vulnerability, Contribution of Working Group II to the Fourth Assessment Report of the Intergovernmental Panel on Climate Change. Cambridge University Press, Cambridge, pp. 7–22.

IPCC, 2012. Summary for policymakers. In: Field, C.B., Barros, V., Stocker, T.F., Qin, D., Dokken, D.J., Ebi, K.L., Mastrandrea, M.D., Mach, K.J., Plattner, G.-K., Allen, S.K., Tignor, M., Midgley, P.M. (Eds.), Managing the Risks of Extreme Events and Disasters to Advance Climate Change Adaptation. A Special Report of Working Groups I and II of the Intergovernmental Panel on Climate Change, Cambridge University Press, Cambridge, UK, and New York, NY, USA, pp. 3–21.

Klinenberg, E., 2002. Heat wave: a social autopsy of disaster. University of Chicago Press, Chicago.

Koppe, C., Jendritzky, G., Pfaff, G., 2003. Die Auswirkungen der Hitzewelle 2003 auf die Gesundheit. In: Klimastatusbericht 2003. Offenbach a.m., pp. 152–162.

Kuttler, W., Dütemeyer, D., Barlang, A.-B., 1997. Klimatologische Untersuchungen in einem potentiellen Beflüftungsareal innerhalb des Stadtgebietes von Köln. Essen (On behalf of the Department of Environmental Protection, City of Cologne).

Luber, G., McGeehin, M., 2008. Climate change and extreme heat events. American Journal of Preventive Medicine 35, 429–435.

Maskrey, A., 1984. Community Based Hazard Mitigation. Disasters Mitigation Program Implementation. Virginia Polytechnic Institute, Virginia.

Meehl, G.A., Tebaldi, C., 2004. More intense, more frequent, and longer lasting heat waves in the 21st Century. Science 305, 994–997.

Pelling, M., 1997. What determines vulnerability to floods; a case study in Georgetown. Guyana. Environmental and Urbanization 9 (1), 203–226.

Pelzing, R., 2008. Erdbeben in Nordrhein-Westfalen. Geologischer Dienst Nordrhein-Westfalen – Landesbetrieb –.

Pinter, N., van der Ploeg, R.R., Schweigert, P., Hoefer, G., 2006. Flood magnification on the River Rhine. Hydrological Processes 20, 147–164.

Re, Munich, 2003. Annual Loss and Damage Statistics. http://www.munichre.com/app_pages/www/@res/pdf/media_relations/press_releases/legacy/pm_2003_12_29_d.pdf.

Robine, J.-M., Cheung, S.L.K., Le Roy, S., Van Oyen, H., Griffiths, C., Michel, J.-P., Herrmann, F.R., 2008. Death toll exceeded 70,000 in Europe during the summer of 2003. Comptes Rendus Biologies 331, 171–178.

Semenza, J.C., Rubin, C.H., Falter, K.H., Selanikio, J.D., Flanders, W.D., Howe, H.L., Wilhelm, J.L., 1996. Heat-related deaths during the July 1995 heat wave in Chicago. New England Journal of Medicine 335, 84–90.

Steinführer, A., Kuhlicke, C., 2007. Social Vulnerability and the 2002 Flood.

Thieken, A.H., Kreibich, H., Müller, M., Merz, B., 2007. Coping with floods: preparedness, response and recovery of flood-affected residents in Germany in 2002. Hydrological Sciences Journal 52, 1016–1037.

Tyagunov, S., Grünthal, G., Wahlström, R., Stempniewski, L., Zschau, J., 2006. Seismic risk mapping for Germany. Natural Hazards Earth System Science 6, 573–586.

UNDINE, 2013. Datengrundlagen zur Einordnung und Bewertung hydrologischer Extreme. Available under: http://undine.bafg.de/servlet/is/8606/.

United Nations, 2005. Hyogo framework for action 2005–2015: building the resilience of nations and communities to disasters. In: World Conference on Disaster Reduction, 18–22 January 2005, Kobe, Hyogo. Available at: http://www.unisdr.org/wcdr/intergover/official-doc/L-docs/Hyogo-framework-for-action english.pdf.

Urban drainage department City of Cologne, 2013. Flood Statistics. Available under: http://www.steb-koeln.de/statistik.html.

Vervuren, P.J.A., Blom, C.W.P.M., de Kroon, H., 2003. Extreme flooding events on the Rhine and the survival and distribution of riparian plant species. Journal of Ecology 91, 135–146.

Vescovi, L., Rebetez, M., Rong, F., 2005. Assessing public health risk due to extremely high temperature events: climate and social parameters. Climate Research 30, 71–78.

Wilhelmi, O.V., Hayden, M.H., 2010. Connecting people and place: a new framework for reducing urban vulnerability to extreme heat. Environmental Research Letters 5, 014021.

Wisner, 2002. Who? What? Where? When? In an Emergency: Notes on Posible Indicators of Vulnerability and Resilience: By Phase of the Disaster Management Cycle and Social Actor. In: Plate, E. (Ed.), Environment and Human Security, Contributions to a workshop in Bonn, pp. 23–25. October 2002, Germany, pp.12/7-12/14..

Wisner, B., Blaikie, P., Cannon, T., Davis, I., 2004. At Risk, second ed. Routledge, London.

Wolf, C., 2002. Urban air pollution and health: an ecological study of chronic rhinosinusitis in Cologne, Germany. Health & Place 8, 129–139.

Chapter 6

Vulnerability to Drought and Heatwave in London: Revealing Institutionally Configured Risk

Zehra Zaidi and Mark Pelling
Department of Geography, King's College London, The Strand, London, United Kingdom

Chapter Outline

6.1 CASE STUDY DESCRIPTION

The case study conducted by King's College London examined two hazard types for the city of London: Drought and Heatwaves. For both hazards, risk management structures at the city and local level were examined as an indication of the level of vulnerability existing in the system.

Assessment of Vulnerability to Natural Hazards. http://dx.doi.org/10.1016/B978-0-12-410528-7.00006-0

6.1.1 Heatwave

For the heatwave hazard, a citywide approach was adopted to understand risk management in London since the procedures and management structures in place to deal with heatwaves hazards are enforced uniformly across the city. This also involved incorporating elements of national risk management since the national Heatwave Plan is implemented by the National Health Service (NHS). Stakeholders and study participants included representatives from the Met Office, national and city level government officers, academics, and emergency management experts. The study established the procedures present for heatwave risk management and explored their impact and robustness in conjunction with practitioners and managers of heatwave risk in London.

The adaptive capacity for heatwaves risk at the local level was examined by using Tanja Wolf's (2009) doctoral research on heatwave vulnerability mapping to identify groups that are most at risk to heatwave hazards in London—i.e., elderly and disabled individuals requiring care—and investigating the incorporation of heatwave risk management in care practices. Given the complexity of care provision in London, two local councils of Waltham Forest and Hackney were selected for field research, featuring four types of care provision: nursing homes, care homes, sheltered housing, and in community care. The results were combined to provide a qualitative assessment of preparedness, risk management, and vulnerability to heatwaves in London.

6.1.2 Drought

For the drought hazard, a cross city sample was taken for both the risk management and local adaptive capacity elements of the study. This is due to the fact that water provision across the capital is fairly equitable and well-regulated, offering a uniform level of service delivery across all local councils. In order to assess the management of drought risk at the city level, water providers, regulating authorities, local government officials, and corporate documentation on drought management procedures were consulted for data collection. Using this data, the structure of drought management practices was established and its robustness explored. The adaptive capacity to drought risk at the local level was conducted by identifying ecological, economic, social, cultural, and institutional aspects of drought risk for local level organizations directly affected by water shortages. Using interviews and expert analysis, the risk management practices of these organizations were explored, and their degree of preparedness and disaster planning and awareness assessed as an indicator of local adaptive capacity to drought risk in London.

6.2 HAZARDS

6.2.1 Heatwave

The irregular nature of heatwaves, including variations manifested in frequency, magnitude, severity, duration, areal extent and onset speed, had led to a lack of consistency in their definition. In general, Meehl and Tebaldi (2004) associate such extreme events with particularly hot sustained temperatures that produce notable impacts on human mortality, regional economies, and ecosystems. Similarly, McGregor et al. (2007) consider heatwaves to be "periods of 'anomalous' heat that generate a societal response". The World Meteorological Society describes heatwaves as periods of warm weather lasting for more than five consecutive days, with temperatures that are five degrees in excess of the average maximum temperature for that region. This criterion for heatwaves is applied in the United Kingdom by the Met Office, with the daytime temperature threshold for most regions in the country set at 30 °C. However, taking into account the urban heat island effect and meteorological conditions across the south east of the country, this average is 2 °C higher for the city of London.

Exposure to extreme heat can cause death if the human body is no longer able to maintain its required temperature through circulation. The ambient temperature at which a body is at risk for heat-related health problems varies with the climatic conditions to which an individual is acclimatized. High levels of humidity and air pollution can compound the effects of heatwaves by hindering respiratory function or the cooling processes of the body through perspiration. Heatwave mortality is predominantly caused by underlying cardiovascular, respiratory, and cerebrovascular conditions (Basu and Samet, 2002).

Vulnerability studies indicate that certain subgroups of the population are more susceptible to extreme heat events. Epidemiological studies examining the relation between heat and health indicate that "people at highest risk of death following heatwaves are over 60, or work in jobs requiring heavy labor, or live in the inner city and/or low-income districts and thus are exposed either to low economic status or higher temperatures or both" (McGregor et al., 2007). Oftentimes, people over the age of 65 years are made more vulnerable to heat stress due to weak thermoregulatory mechanisms, compounded by chronic dehydration, the side effects of medication, and illnesses affecting functions that regulate body temperature (Worfolk, 2000). A positive correlation has also been observed between those that are elderly and living in a care home, and increased vulnerability to heat-related illness and death (Faunt et al., 1995). This may be a result of high indoor temperatures in institutions such as hospitals and care homes, as demonstrated during the during the 2003 heatwave in London (Newton, 2005).

The impact of heatwaves is generally measured using mortality and morbidity numbers, although this is not necessarily a good indicator of heatwave vulnerability in society. As McGregor et al. (2007) indicate, "systems for life support (energy, drinking water and sanitation, food distribution), social development (health, education, community development and social support), innovation (cultural and intellectual services including the media), communication (transport, telephone and IT networks), social control (policing and regulatory functions) and the economy (private markets and financial services)", all play a role in determining the severity of the impact of a heatwave. Therefore, infrastructure and institutional response play an important role in mediating the risk and impact of extreme heat hazards, as do the social determinants of vulnerability.

London was one of the worst affected European cities in terms of mortality numbers for the 2003 heatwave. The Met Office predicts a rise in the incidence of extreme hot weather periods over the coming years.

6.2.2 Drought

Most conceptions of drought are relative and vary in scope due to the diverging intensity, duration, and spatial extent of the phenomenon. In general, scarcity of moisture—whether it is in the form of rainfall, soil moisture, or groundwater—is evaluated in relation to the demand for that moisture in a particular system. At their simplest, droughts can be viewed as a simple equation of water demand outstripping supply. From a social perspective, a drought is seen to occur when such a shortfall between supply and demand causes human hardship (Heathcote, 1991). According to the UK Environment Agency, a drought takes place when "a period of low rainfall leads to a shortage of water" (Environment Agency, 2007).

In the examination of droughts, precipitation is considered as the primary factor controlling the creation and persistence of drought conditions, but evapotranspiration is also an important variable. However, most research into the general effects of drought relies on the most readily available data—rainfall statistics, and suggests that there is a strong correlation between meteorological droughts (annual rainfall falling in the lowest 10% of values on record) and the occurrence of agricultural, socioeconomic, or urban droughts (Bond et al., 2008; Mpelasoka et al., 2008).

The use of a meteorological framework for the understanding of droughts reveals itself as problematic when it becomes apparent that human management of resources can either reduce or enhance the risk of drought occurring by reducing or increasing the provision and demand for water. Different thresholds of water need and different management systems create different potentials for drought occurrence. Both are a product of social and technological variables.

Hewitt (1997) reports that throughout the world, drought ranks first among natural disasters in numbers of persons directly affected. In an urban context, rising standards of living and increasing populations mean that water demand in cities is rocketing (Fitzhugh and Richter, 2004). According to the UK Environment Agency, London is the driest capital city in all of Europe (GLA, 2007). In cities, the impact of drought depends largely on societal vulnerability at the time when the drought occurs. Rising population and water demand in London, combined with an aging supply system, have led to pressures on the water cycle and recent climatic changes have resulted in an increase in the incidence of droughts in the capital (GLA, 2006). The Environment Agency believes that there is a high probability that coming summers in London are hot and dry, which might result in more severe droughts occurring than have been experienced in the past 100 years.

6.3 VULNERABILITY ASSESSMENT METHODS

This section provides an overview of the structure of heatwave and drought risk management practices in London that shaped the criteria for selection of the conceptual framework and methodology of the London case study. Following this contextual background, an account of the derivational process involved in index development is outlined.

6.3.1 Heatwave

Following the large number of deaths in the heatwave of 2003, the Department of Health introduced a National Heatwave Plan for the United Kingdom in 2004. The plan sets out the arrangements that will apply and the actions required in advance of, and during a heatwave. It is revised on an annual basis and disseminated by the Department of Health to strategic health authorities and local health authorities for implementation. At the local level, health provision associations, Primary Care Trusts (PCT), General Practitioners (GP), lead nurses, directors of social services, directors of housing and planning, professional bodies, and care associations are some of the organizations and individuals involved in the implementation of the guidelines set in the Heatwave Plan.

Heatwaves are short term and relatively infrequent events in London. This makes it challenging to evaluate management preparedness and response to them. They also affect a diverse group of individuals, and impact differently across geographical, social, and economic scales. In the United Kingdom, heatwave impacts are primarily measured in terms of additional morbidity and mortality, with impacts seen as a consequence of hazard and vulnerability together with the limits or failures in risk management. A reliance on only mortality rates and socioeconomic data as the primary measure of vulnerability can discount the capacity of social behavior and institutional change in shaping risk, leading to vulnerability appearing as a static measure rather than a continuously changing social and physical capacity to withstand extreme heat. For example, Hajat et al. (2007) and Bell et al. (2008) differentiate vulnerability to heat-related mortality along the lines of gender, age, and economic access, but in neither case are key data on the behavior of the vulnerable and their care providers, their use of internal space, and the performance of organizations and institutions that regulate their access to information and support considered in the assessment. This is a critical gap for the MOVE project to examine.

In London, efforts have been made to map vulnerability to heatwaves, most notably by Wolf (2009). Similar to efforts in other countries, this study incorporates GIS and socioeconomic data to present areas and groups most at risk in London. However, although the Wolf (2009) study provided a preexisting "measure" of vulnerability for the MOVE London case study, it also served to highlight the methodological and conceptual limitations of vulnerability studies that fail to incorporate the effect of social practices and institutional risk management. When addressing the issue of vulnerability, nearly all the reviewed theoretical literature acknowledges that human behavior can/does lead to adaptation in both the long and short term, and that this process of social adaptation can be facilitated by better health warning systems, heatwave preparedness or action plans, and better information on who is most affected (Hajat et al., 2006; Conti et al., 2005; Stafoggia et al., 2006). The sets of vulnerability identified in the Wolf study were temporally static and did not reflect the role of adaptive practices, both at the public policy and individual level, in constantly shifting vulnerability.

This critique is reinforced by research undertaken by Abrahamson et al. (2008), who found that elderly people's perceptions and knowledge of heat-related health risks, and protective behavior patterns were central in shaping vulnerability in two UK cities. In order to construct a more dynamic and accurate assessment of vulnerability in London, the concept of adaptive capacity was incorporated into a calculation of vulnerability in the King's College London (KCL) study.

Similarly, it is important to note that few studies have thus far incorporated institutional capacity and infrastructural vulnerability in overall assessments of vulnerability to heatwaves. There is also a lack of systematic studies into the relationships between key service providers and their public sector risk regulators. Abrahamson and Raine (2009) explore the perceptions of frontline staff in state medical care facilities, social services, and volunteer groups on the feasibility of implementing the UK Heatwave Plan for the elderly. As observed in their research, most health care workers and voluntary staff servicing the elderly in London had never heard of the Heatwave Plan prior to the study. These findings highlight the mutually constitutive role played by policy and action plans, and the importance of behavior and practices of implementing agencies, actors, and those at risk in shaping overall vulnerability to heatwaves.

In addition, heatwave management requires coordination across policy sectors and between individual service providers in particular locations. Evaluating the risk management component of vulnerability therefore relies on a subset of indicators that can encompass a wide range of organizational and social factors. This is needed to allow the capture of the full range of potential policy mixes and associated actors from government to civil society and the private sector in the provision of risk management. The relationship between key service providers, public sector regulators, and those most vulnerable to heatwaves needs to be examined in greater detail in order to strengthen institutional capacity, and support greater relevance between policy measures and social adaptive behavior in the context of heatwaves. The ways in which different care provision contexts shape delivery of the London Heatwave Plan from the perspective of different stakeholders including the vulnerable elderly is a critical gap for the MOVE project to examine.

6.3.2 Drought

The water provision industry in London has a complex history, which is complicated by the juxtaposed roles played by the public and private sectors in regulation and distribution services. Currently, the primary water supply company in the London metropolitan area is Thames Water Utilities Ltd. Its water supply and sewage operations are supplemented by three other supply-only companies, namely Sutton and East Surrey, Three Valleys Water, and Essex and Suffolk Water. These companies represent the public face of the water industry in London. It is their duty to develop and maintain efficient and economical systems of water service provision. In turn, customers have a statutory right that guarantees the appointed water company will connect them to the network and provide them with a constant water service.

In addition, the responsibility for the development of policy and legislative frameworks, control of pollution and environmental regulation, and other regulatory enforcement is shared by the Environment Agency, Department for Environment, Food and Rural Affairs (DEFRA), Office of Water Services (OFWAT), District Health Authorities, local authorities, and other public institutions. The boundaries between the respective responsibilities of each agency and department listed above are, at times, vague. Indeed there is no single department with legal responsibility for drought or heatwave risk management as defined under the Civil Contingencies Act 2004. Over the years, changes to the roles of the different bodies have diminished the public understanding of who is responsible for what task. Likewise, the different interpretations of responsibility have led to reluctance toward incurring costs in the absence of clear funding procedures, particularly for storm water and urban drainage. This has impacted the management and regulation of drought adaptation to a certain degree (GLA, 2007).

The Environment Agency has recently published a *Consultation on Identifying Areas of Water Stress* that highlights London as an area of serious water stress. During most summers, there is sufficient water in the rivers Thames and Lee to meet London's demand for water. It is periods of low rainfall that threaten the security of supply. Low rainfall over the winter months limits the refill of groundwater stores, which in turn leads to low river flows in the following spring and summer. Typically it takes two winters of below average rainfall to precipitate drought actions.

The Water Act 2003 requires all water companies to have sound drought plans in place so that they can continue to supply water to their customers in times of drought or when sources are depleted. Drought plans are an important element of effective water resources management in London and detail the operational steps that must be taken as drought progresses. The different stages of the plan and the measures that are to be implemented need to be linked to specific drought "triggers" during the escalation and de-escalation of a drought. Triggers should be identified in advance, and the crossing of a trigger should prompt a company to initiate predetermined actions or to move to the next stage of drought management. By planning these actions in advance, there is time to consider potential impacts and mitigation measures. As part of such measures, companies can implement actions to manage demand, such as introducing customer restrictions, or to temporarily increase supplies, for example, by applying for drought permits or drought orders.

Efforts toward ensuring that all residents in the Greater London area enjoy equal and reliable access to water during periods of water scarcity have minimized the potential social impact of drought in the city. Over the long term, the brunt of water management improvement efforts remain focused on providing engineering solutions to prevent loss through leakages and expand supply sources since institutional practice and learning is generally well developed. Critics argue that the privatized nature of the water sector in London means that there is little incentive for companies to encourage demand management.

However, one reason for this emphasis on supply management and engineering is the large quantity of water loss experienced in the capital. Large parts of London's water supply network date back to the Victorian era; over 60% of the network is pre-1900. It is the aged infrastructure that leads to most parts of London having the highest levels of mains leakage in the United Kingdom. While only 70% of Thames Water's customers reside in London, 85% of Thames Water's distribution losses occurred in London. Alternatively, Thames Water accounts for 92% of all distribution losses in London, yet only serves 76% of London's population. Leakage in London ranges from 85 to 260l per property per day. Yet having a system with very low leakage levels would prove too costly. Thus, there is some tolerated level of leakage. The Government, regulators, and water companies accept the economic level of leakage (ELL) as the preferred tolerated level of leakage. The ELL is the point at which any further effort to reduce leakage would not be cost-effective; or in other words, the point at which it would cost more to reduce leakage further than it would be to produce water from an alternative source.

As a result of such economic and physical constraints, and the relative lack of drought experience in London, water management and drought risk reduction remain focused on engineering led supply management rather than a transformation of social practices or demand management. The construction of additional reservoirs, storage, and desalination plants is the primary means of reducing drought risk in London. This has resulted in drought management being restricted to public and corporate guidelines at the service delivery level, with very little impact on local level practices within the city.

6.3.3 Index Development

After an examination of the overall context of vulnerability for heatwave and drought hazards, a common index was considered to be sufficient for capturing both sets of vulnerability in London. Measuring vulnerability to heatwaves and drought requires the assessment of several discrete properties. These are identified in the MOVE conceptual framework. The KCL index seeks only to capture that part of the vulnerability dynamic that is affected by the capacity and character of hazard risk reduction and response. These are highlighted in the Risk Governance and Adaptation sections of MOVE framework. Clearly governance and adaptation do not stand alone and the KCL index also includes capacity for risk management systems to obtain information about Hazard and Vulnerability both in anticipation of and as a measure of management fit and feedback from management decisions and actions on risk.

The KCL index is based on the risk management index (RMI) (Cardona, 2005) but oriented toward the nonstructural impacts of hazards and operating at the scale of a mega-city. As with the RMI, this is a composite indicator. Greater emphasis is placed on adaptive management than in the original RMI. It measures disaster risk management in terms of the perceived performance of public policy and local adaptive capacity for four fields: risk identification, risk reduction, disaster management, and adaptive governance. Each policy field is evaluated based on the benchmarking of a set of subindicators that reflect performance targets associated with the effectiveness of disaster management activities. The participation of external experts as well as disaster managers is incorporated to minimize bias.

In order to capture the layers of risk management operating within the city scale, the index is applied at two subdimensions in the form of the RMI and the local adaptive capacity (LAC). The RMI evaluates hazard risk reduction and response at the level of public policy and risk management practitioners. The RMI is complemented by the LAC, a composite indicator that measures adaptive capacity at the level of individual organizations from small businesses to local authorities. Emphasis in the LAC is on the capacity to act as framed by overarching public institutions as well as the capacity of individual organizations. Similar to the RMI, the LAC index is composed of indicators for risk identification, risk reduction, disaster management, and adaptive governance. Both indexes are intended for generic use across any area at risk from heatwaves or drought.

The RMI and LAC are a participatory risk index. This means that the methodology used to derive the index was itself an opportunity for respondents to reflect on their own practice. The outcome is an index based on generic themes but refined through interviews with London-based respondents to emphasize aspects of risk management most germane to London. The index was built in three steps. First a basic structure for the index was derived deductively. At this first stage, the deductively generated KCL index had a simple structure with five component indicators, each populated with input variables, but with no weightings attributed. Stage I was taken from the generic RMI index developed by Cardona (2005) with one additional field "Adaptive management" added to highlight its importance. No input variables were predefined for this component indicator. The Stage I index was offered to respondents for refinement and to gauge their appropriateness. Interviews with key individuals involved in managing and evaluating risk to drought and heatwaves in London were used in conjunction with a literature review to provide a basis for assessing the validity of the indicators presented in the index. Tables 6.1 and 6.2 provide a sample of the areas of practice from which respondents were recruited and the specific roles of individual respondents with respect to heatwave and drought management, respectively. This table also serves to identify any gaps that might bias the remodeling of the generic RMI to measure heatwave risk. Respondents were asked both to:

- Reflect on the suitability of the generic index structure or component indexes, and input variables for heatwave risk; and
- Provide background information on the management of heatwave risk in London.

TABLE 6.1 Heatwave Respondent Groups and Individual Roles (Risk Management Index)

Respondent Group	Individual Respondent Roles
Hazard science community	Met office climate modeler
Heatwave policy practitioners	NHS heatwave plan coordinator
Heatwave policy practitioners	Heatwave plan regulator in health protection agency
Local government risk management	London local council emergency planner
Local government risk management	London Resilience Team
Academic expert	Epidemiological advisor to heatwave plan (UK)
Academic expert	Researcher on impact of heatwave plan (UK)
Academic expert	Gerontology and older people in London expert
Voluntary social care expert	Director of "Help the Aged" London

TABLE 6.2 Drought Respondent Groups and Individual Roles (LAC)

Respondent Group	Individual Respondent Roles
Parks and conservation managers	Species and Estate manager, Royal Kew Gardens
Garden centers and plant stockists	Managers of various garden centers
Garden allotment holders	Head of London Allotments Association
Local government managers	Waltham Forest Tree manager
Health and policy practitioners	Health Protection Agency (drought impact officer)
Voluntary organization	Member of "The Tree Service" in London

The latter information was not used to structure the index, but was useful in preparing for the Stage III assessment and in helping to identify participants for the Stage III assessment.

The index structure was broadly supported. The only change in structure was to remove financial protection and governance, and replace it with adaptive governance as a standalone category. Rather, the financial aspect of management capacity is better reflected and integrated across all other component indicators and will appear at the level of input variables.

Finally a small group of experts were invited to provide relative values for individual input variables in the index as part of the validation process. The advantage of a compound index at this stage is that it allows relative assessment across a range or risk management activities for any one place. This made it easier for respondents to place values on input variables. Ideally the index methodology can also be applied to other places so that comparative assessment can be provided. At a minimum, the third stage of analysis can be applied in this way. The intended final outputs of the indexing methodology involved increased sensitization among key respondents that were provided a space for reflection on practice, as well as the derivation of an index value and values for index components and input variables.

The following section will assess the relevance of each input variable in light of the experience of heatwave and drought risk management in London.

6.4 INDICATORS

This section provides a description of the indicators and subindicators used in the common index framework for both drought and heatwave hazards at the local and risk management level. Each sub indicator has a set of attached input variables that differ across hazard types and reflect the specific characteristics of each case study at both scales of application.

6.4.1 Discussion of Indicators

6.4.1.1 Risk Identification

Risk identification is an important component of hazard management. Prior knowledge of the frequency, duration, and impact of the hazard is essential to effective risk management and vulnerability reduction. The response time for mobilizing disaster response initiatives and procedures needs to be quick in order to manage and limit the impact of extreme weather events such as heatwaves and drought. The availability of risk identification mechanisms and early warning systems is therefore an integral part of an effective heatwave management system. The following sub indicators were considered to be essential in capturing all the dimensions of risk identification for both hazard types:

1. Systematic inventory of disasters and losses:
   The existence of data on the historical frequency of hazards and their impact forms the basis of a robust disaster management system. Meteorological

records of daily ambient temperatures or precipitation provide a record of weather and climatic patterns, and assist in their classification as a potential risk to the population. Losses during a hot weather or drought event are generally calculated in terms of their impact on human mortality and morbidity, although environmental and ecological losses are also important in calculating drought vulnerability. Observed trends in mortality and morbidity rates during heatwaves can be compared with data from preceding years to present an overview of the impact on human health during a heatwave.

2. Risk monitoring and forecasting:
   The presence of scientific mechanisms for predicting and monitoring the meteorological occurrence of extreme heat or drought events is an important aspect of effective risk management. Also important is the existence of a formal and predefined system of risk communication, with both climate scientists and policy makers clear on a common framework of risk classification for hazard events.
3. Vulnerability and risk assessment:
   The identification and measurement of vulnerable individuals and groups in a particular setting is key to an effective risk management system. Vulnerability to hazards can be dynamic, and is seen to vary across social groups, physical space, and different scales. An ongoing review of vulnerable populations is necessary for an effective risk management system; however, general trends observed in historical data provide an overall picture of those most at risk to hazards.
4. Dissemination of information on risk and response measures to risk managers, at risk groups, and care providers:
   Discussions with experts and practitioners of risk management in London highlighted the importance of disseminating public information before and during an extreme heat or drought event. In most cases, risk to such hazards can be reduced through increased alertness and taking simple "common sense" measures. Since drought and heatwave vulnerability are mainly measured by their human impact, providing basic information on risk identification can make it relatively easy to alleviate risk and exposure to such hazards through encouraging minor behavioral changes, and makes public information and community awareness critical in hazard risk management. This input variable is important because it affects the degree of preparedness and risk identification that exists at the local level.

6.4.1.2 Risk Reduction

Risk reduction measures are preevent management activities designed to either directly enforce or empower local actors to contain local human vulnerability and hazard, and enhance adaptive capacity and actions in the long and short term. Several steps can be taken to address heatwave risk and the following input variables are assessed for their relevance in evaluating such risk management systems.

1. Hazard and vulnerability considered in land use and urban planning:
Hazard and vulnerability consideration in the urban planning and the construction of new infrastructure can greatly reduce vulnerability resulting from the built environment. Although such actions have limited impact in mitigating hazard-related risk in the context of cities such as London where changes in existing land use and urban design are financially untenable, management of land use and urban planning can be a key input variable for assessing long-term risk reduction in other contexts.
2. Policy and financial support for alleviating risk or retrofitting buildings in high risk areas
The infrastructure of public buildings and assets such as community spaces and transport networks are often not easily adapted to an increasing risk of climate-related hazards. This includes the network of hospitals and residential care buildings in which those that are already vulnerable reside, thereby amplifying the risk posed by hot weather. The provision of financial support or policy initiatives to undertake adaption measures to make public infrastructure more resilient to heatwaves or drought events can be critical in reducing vulnerability.
3. Public education on risk
The predominantly human impact of heatwaves and drought hazards makes behavioral changes important in reducing vulnerability. Widespread dissemination of information on drought and hazard risk is needed to influence the diverse range of social, financial, ecological, cultural, and institutional activities affected. Public guidelines on adopting alternative practices and updating infrastructure can assist in the long-term resilience to hazard events. Early warning systems are a potent component of public information since heatwaves and drought are both preceded by days of hot or dry weather. This allows time for issuing health warnings and information before the possible onset of extreme temperatures and water scarcity.
4. Regulation of safety measures
Monitoring and regulation of safety processes and procedures in place to mitigate and reduce vulnerability to climate change hazards such as drought and heatwaves need to be carried out in order to ensure relevant and robust risk management.

6.4.1.3 Disaster Response

1. Emergency response plan or procedures
The creation of a formal emergency plan to deal with extreme weather events can greatly improve the risk management capacity of a particular society. Such plans can exist at the national or subnational level, and outline the procedures and protocol to be followed by a number of different agencies in order to effectively manage risk during hazard events. A heatwave or

drought plan should identify individuals and groups mostly at risk and set out a method for communicating risk both between emergency planners and practitioners, and to the public at large.

2. Availability of disaster response equipment, facilities, and skills
   Adequate emergency medical facilities and staff need to be present in a system in order to effectively manage health risks during a hazard event. In addition to this, a network of social care workers and staff looking after vulnerable population groups can assist in the provision of assistance and advice to individuals at risk during such periods. In most cases, the presence of a functioning health care system is sufficient infrastructure for the management of heatwaves. In the case of drought, facilities can include reserve supplies and storage of water. Overall, this is a useful input variable for assessing risk management capacity.
3. Community and private sector preparedness and training
   Community and private sector preparedness serve to complement disaster response efforts by addressing vulnerability in households, work spaces, and in communities.
4. Disaster response training
   Similar to community awareness, but perhaps of greater consequence, is the need for practitioners and frontline staff responsible for dealing with people vulnerable to climate hazards to gain adequate training and education regarding risk management procedures and processes for an effective heatwave management system.

6.4.1.4 Adaptive Governance

Adaptive governance measures the degree to which the existing system has inbuilt mechanisms for adaptive flexibility (change within existing limits of practice) and adaptive reform (a timely changing of the limits). Not only is it necessary to assess the current status of the risk management system, it is also important to appraise room for change within the current institutional framework, and the ability to reform the current institutional framework should it prove wanting. The following subindicators gauge adaptive capacity by examining the technical and also some element of the political systems influencing risk management.

1. Horizon scanning for unexpected risks
   The ability of risk management systems to continually assess and reexamine the conditions of vulnerability and risk is an important indicator for a flexible regime with a capacity to absorb and predict disaster risk.
2. Ability of policy and implementing agencies to reflect on practice outcomes
   This input variable highlights the adaptive behavior and institutional learning capacity present in a risk management system. It is important to continuously update and improve the practices and techniques involved in monitoring and managing heatwaves due to their dynamic impact and shifting vulnerability in society. Also, risk management of heatwaves relies largely on human actions and modification of institutional practices and individual behavior

rather than alterations in infrastructural or environmental management. This makes the testing and alteration of interinstitutional response and practice an important factor in improving and adapting heatwave management systems to better address vulnerability and risk.

3. Support for experiments in risk reduction and response
   This indicator measures the degree of flexibility within a management system to support adaptive learning and experimentation in its practice and policy.
4. Availability of funds for flexible vulnerability management
   Budget allocation and mobilization is required for creating and revising risk management systems, and to strengthen coordination and training between different departments involved in its implementation.

6.5 RESULTS AND VALIDATION

6.5.1 Heatwaves (RMI)

The application of the RMI index to heatwave risk management at the national, city, and subcity level returned consistent results from a variety of management practitioners and expert commentators. Overall, the index revealed a strong structure for heatwave risk identification in London. A positive result for the subindicators "systematic inventory of past events" and "risk monitoring and forecasting" was in part due to the high level of climatic monitoring and forecasting practiced by the Meteorological Office at the national scale, and its proactive role in disseminating information regarding extreme weather events. The system also performed well in the vulnerability and risk assessment category, although vulnerability was measured within the national framework solely as an outcome of medical factors without taking into account social variables. A new system for categorizing cause of death at hospitals had also allowed for greater accuracy in calculating heatwave mortality. Procedures for the dissemination of information on heatwave risk were well organized and clearly outlined, although their organization within the Health Department meant that the dissemination mechanisms were deployed through the NHS and did not involve social services.

In the risk reduction category, the existence of old infrastructure and building stock in London, the high costs associated with renovating and retrofitting, and the relatively small scope for new construction in the city meant that the subindicators for hazard and vulnerability considered in land use and planning, policy, and financial support for alleviating risk or retrofitting residential and working space in high risk areas, public education on risk reduction, and regulation of safety measures had a low level of applicability to London. It is important to note that existing building codes and regulations for new construction put more stress on insulating buildings against cold weather, which often has the inverse effect of making infrastructure more vulnerable to extreme hot weather. On a city level, the Mayor of London's office has recently conducted studies into the urban heat island effect in London and has acknowledged the need to incorporate land use strategies and urban planning that offset the impact of this urban phenomenon.

The existence of the National Heatwave Plan ensures well-developed procedures for heatwave disaster response in London. However, the failure to involve agencies and departments other than the health and emergency services has resulted in limited awareness of the plan outside of the health management system. In addition, although there is a good availability of disaster response equipment, facilities and medical skills to manage heatwave disaster response, limited outreach and information on heatwave risk within the community and private sector means that at-risk groups and carers are often unprepared and untrained to identify and respond to emergencies resulting from extreme hot weather. This gap in disaster response is reflected in the persisting proportion of heat-related mortality and morbidity in elderly groups living in the community or in care homes across London (and was also validated by the results of the heatwave LAC).

Currently, the national Heatwave Plan is revised on an annual basis and its effectiveness is independently reviewed by the Health Protection Agency every two years or immediately after an episode of extreme hot weather. The presence of such mechanisms results in a high degree of adaptive governance by providing platforms for policy makers and implementing agencies to reflect on practice outcomes. However, although the Department of Health is supporting research for the incorporation of social indicators in framing heatwave vulnerability, the continued understanding of heatwave disaster management as a medically driven response rather than both a medical and social issue has resulted in the exclusion of social service practitioners and care workers from having a role in the formulation and implementation of the plan.

6.5.2 Heatwaves (LAC)

The application of the index to assess local adaptive capacity revealed varied results. This was in part due to the inconsistent standards of care provision across private, public, and voluntary organizations, as well as the consideration of four different levels of care provision ranging from nursing homes to in community care. Most care providers had an awareness of heatwaves posing a potentially serious risk for elderly and disabled people. Informal records and recollections of hot weather events and their impacts were found in all organizations, but only a few private homes had facilities to monitor and record daily indoor temperatures. Risk monitoring and forecasting were generally strong due to the dissemination of health warnings by local PCTs to care providers via email. However, this trend weakened out toward the sheltered housing and in community care sections since information usually disseminated during weekly or biweekly meeting with local council care manager, which were too far apart to effectively inform on risk of hot weather events that occur for short periods of time. In such cases, awareness depended on a carer's access to Internet and personal habits of checking email or other advisory sites. Therefore, more needs to be done to ensure an effective chain of communication between health workers, PCTs, and care providers.

Risk reduction practices were limited in scope across all organizations. Most care institutions existed in old infrastructure. Even where the buildings were relatively new or brand new, there had been no consideration for hot weather resistance during design and construction. In fact, safety measures and cold weather proofing meant most homes had windows that could not be opened and suffered from poor ventilation. There was also little financial support for alleviating risk or retrofitting residential and working space in high risk areas, with funds often limited to the provision of fans. Public education on risk reduction and regulation of safety measures was also low. Only one organization had an evacuation procedure in place for hot weather, and public information campaigns only took place during summers and were more about response rather than risk reduction. The risk reduction practices of community carers were limited to informing residents on basic precautions against hot weather during such events.

Most care practitioners were not formally aware of the national heatwave plan. However, they were aware of an advisory information service by the PCT that would warn them of such events. Where there was a good flow of information down from the PCTs to carers, an effort was made by carers to print and disseminate leaflets issued by the health department within their organization. However, those with limited access to NHS information or Internet access relied largely on information from watching local news on TV. Also, the dissemination of information by the PCT did not guarantee adequate response. Some care managers, mostly in public and voluntary institutions, did not appear to have adequate knowledge of treating heatwave vulnerability, with most focused on hydrating patients rather than the need to regulate fluids and temperature. Although NHS directives pin pointed those with respiratory and heart illnesses as more vulnerable, most care managers did not appear to be aware of how the two were linked to risks from high ambient temperatures.

Although there was no disaster response training specifically for heatwaves in care homes, the daily monitoring of patients and the ongoing assessment of their medical conditions that formed a part of everyday management practices in nursing and care homes ensured daily care interventions and monitoring responses for dealing with heat-related problems. For these types of organizations, the quality of preparedness and response to heatwave risk was directly linked with the quality of overall care in the organization itself. In contrast, risk management was relatively weak in sheltered housing and in community care, irrespective of the quality of staff or care, since these types of care provision did not involve daily contact with residents or the proactive medical monitoring of elderly or disabled patients. The relatively quick onset of heatwaves, shifting conditions of vulnerability, and weaker levels of communication and individual contact meant that unless residents had strong social networks of support, vulnerability was higher in these groups.

There was little indication of adaptive governance at the local level, and most organizations and individual carers were not involved in any reflective or learning activities. However, most were open to following advice and instructions issued by the local PCT on adapting practices and routines, thus highlighting the importance of adaptive learning at the level of risk management in the health department.

6.5.3 Drought (RMI)

As mentioned in the methodology section, the application of the RMI to drought management in London was applied using extensive literature reviews due to the unavailability of drought managers as well as the detailed information available in the plans and literature of the water companies and other drought management agencies. In all the categories of risk identification—i.e. systematic inventory of past impacts, risk monitoring and forecasting, vulnerability and risk assessment, and dissemination of information on risk and response measures to at-risk groups, providers, responders—the index revealed positive results with the exception of the last subindicator, where targeted public information campaigns for businesses and affected individuals was a little weak.

Practices for risk reduction were primarily focused on supply side solutions. This reflected the engineering focused approach of the water companies, which are interested in expanding supply mechanisms rather than limiting demand. However, strong regulations by government require the metering of all new properties and the consideration of water supply in future land use and planning. In addition, financial support for the free retrofitting of residential stock and businesses with meters has also encouraged regulation of demand, though with limited effect. Public education on risk reduction and reduction of water usage remains very low with little effort made to curb water consumption.

Water companies are legally bound to produce emergency response plans and procedures for drought management. These plans rely on restricting usage and demand only as a last resort and instead focus on expansion of supply through engineering solutions and limiting leakages. This has affected the availability and type of disaster response equipment, facilities, and skills used to manage drought, with the construction of desalination plants and creation of storage reservoirs given primary precedence. Limitation to no community and private sector preparedness and training exists for drought management, as is the case with disaster response training.

The legal requirements imposed on water companies require them to annually revise their drought plans to reflect changing conditions in water supply and demand. However, this exercise is aimed at improving supply and limiting leakages. Therefore, although a strong level of adaptive governance—including horizon scanning for unexpected risks, ability of policy and implementing agencies to reflect on practice outcomes, support for experiments in risk reduction and response, and availability of funds for flexible vulnerability management—exists, it is limited to the engineering sphere.

6.5.4 Drought (LAC)

The application of the LAC index to drought risk management in London revealed consistent trends, although variation existed across the different groups of practitioners interviewed.

There was little evidence of formal risk identification practices across any of the respondents involved in the study. No systematic inventory of past events was maintained in any organization except for Kew Gardens, which had in-house facilities for monitoring precipitation and recording weather patterns. However, experiential memories in staff served to provide informal records of such events. Risk monitoring and forecasting was provided by the Met Office and gardening information resources, however these were not monitored as part of management practices except for allotment holders since they were the group with least access to unlimited water supplies and hence more reliant on rainfall for water provision. Every group demonstrated the existence of vulnerability and risk assessment procedures, with an acute awareness of stock vulnerability and potential loss. There was also a good network for communication and dissemination of information on drought risk and response measures, but this was provided by other local level practitioners and gardening enthusiasts rather than water companies or the environment agency.

There was a mixed result for risk reduction practices. Gardening centers were acutely aware of changing climatic conditions and the need for water conservation and had responded by changing their stock to more hardy and drought resistant varieties. Small infrastructural changes to assist in water collection and limit water wastage had also been carried out by garden centers. However, there was relatively little effort to reduce risk by allotment holders since most engaged in gardening as a hobby and did not mind losing stock due to a drought event. In some instances, individual allotment holders or organizations had made provisions for small scale water storage, but this was due to an interest in reducing conflict over sharing common water supply rather than as a drought prevention mechanism. In the case of the Royal Parks such as Kew Gardens, no effort had been made to reduce risk to drought since a steady supply of water was guaranteed to the organization even in times of drought and the cost of retrofitting buildings with more efficient water usage systems was too high in contrast to just paying for the water supply.

There were no disaster response strategies or procedures in place on a formal level. This was because garden centers, parks, and tree services are exempt from drought orders. Even in the case of extreme events, the availability of water tankers from private suppliers would be used to fill the gap. The only group directly impacted by drought conditions would be allotment holders but as mentioned above, since there was no financial or ecological loss associated with loss of crops and plants, not many were overly concerned since plants could be replaced or regrown. Individual water stores provided some potential relief measures.

Low levels of disaster management practices meant that there were correspondingly low levels of adaptive governance or learning at the local level. Although horizon scanning for unexpected risks formed a part of most gardening practices, it was limited to short-term fluctuations in weather rather than a concern with climatic conditions or trends. In the case of Kew and garden centers, there was some effort to reflect on practice outcomes (usually spearheaded by motivated individuals), but low levels of organizational support for experiments in risk reduction and response, and limited availability of funds for flexible vulnerability management resulted in little impact of such practices.

Overall, a high level of awareness of drought risk was demonstrated among respondents at the local level but this resulted in little action to change practices unless there were financial incentives or consequences (as for garden centers) to such adaptive measures. Some evidence of long-term adaptation and learning (stock change, water conservation) was found, but this was of limited scope and was not relevant during disaster management periods since most organizations were not affected by drought orders.

The process and results of the index were validated through the involvement of participants and expert commentators that were involved in the revision and outcomes of the research.

6.6 DISCUSSION

6.6.1 Heatwave

Overall, the data revealed that heatwaves are not considered to be a high priority threat for London. Both at the risk management and local level, there was more emphasis placed on preparedness and adaptation to cold weather and other hazards such as flooding. This was exacerbated by a variation and lack of consistency in care provision standards at the local level. While certain institutions seemed well prepared for heatwave risk, poor training of nonmedical staff and carers in recognizing and addressing heatwave risk appeared to be a problem. Overall, the level of risk management was largely determined by quality of staff and training, individual initiative by manager or council, or financial assistance available to an organization.

At the macro level, the application of the RMI and LAC index to heatwave vulnerability in London revealed fractures in risk management practices that exacerbate existing vulnerabilities across the city. Despite the existence of a coherent and well-revised Heatwave Plan, risk reduction guidance and disaster management procedures do not always filter down to the local level. Also, the perception of heatwave vulnerability as a medical issue has led to risk management efforts focused on preventative measures within the health sector. However, as the LAC data reveal, vulnerable individuals (especially those living in the community) are often already exposed to the risk by the time they come into contact with health services. In order to alleviate and reduce risk, social workers, carers, and community members need to have a greater role in preventing and reducing exposure to heatwave risk for those most vulnerable. Therefore, a potential gap in risk identification and management practices is the understanding of heatwave vulnerability solely as a medical issue. The inclusion of social indicators as contributing factors to heatwave vulnerability (such as consideration of individuals living alone) points to the need for social services to be involved in the risk management structure. This point is reinforced by Tanja Wolf's work (Wolf, 2009) which successfully maps vulnerability as a composite of both social and physical indicators but nonetheless fails to capture the dynamic nature of this vulnerability which shifts according to changing social circumstances. In terms of the MOVE framework, this demonstrates that although vulnerability mapping using demographic and physical data provides a good starting point for assessing existing vulnerability in a system, it is nonetheless a static snapshot of vulnerability, which is generally (but especially in the case of heatwaves) a dynamic phenomenon.

The role of adaptive governance and institutional learning is an important way of dealing with shifting vulnerability. However, as seen in the London case study, even though formal procedures for revision and learning were explicitly built in to the Heatwave management Plan, the presence of such mechanisms did not in itself guarantee true learning. This appears to be because of the unwillingness on the part of practitioners to involve new actors (and the existing tensions and lack of coordination between social and health departments in the United Kingdom).

The application and use of the framework proved useful for highlighting areas of weakness in the overall structure of risk management for heatwaves in London. At the local level, most of the organizations and individuals involved in data collection found the index to be a useful tool in providing an overall assessment of their risk management practices. In some cases, organizations expressed an interest in the comparative management practices of other organizations and the need for knowledge sharing across care providers. The use of a common index led to exchange of information between respondents, reflection on practice and policies, and increased focus on heatwave management in the local organizations involved in the data collection.

However, it must be noted that recent funding cuts, government restructuring, and the dissolution of PCTs and organizations such as the HPA that played a very important role in the risk management process will have a severe impact of heatwave risk management in London.

6.6.2 Drought

The lack of any immediate historical experiences of drought in London limits an assessment of existing emergency drought measures. Overall, an equitable supply and water provision system means low levels of differentiation or social vulnerability at the local level. However, persistently high levels of leakages specific to London's aging infrastructure have resulted in a drought management approach focused on engineering solutions and supply management. Efforts remain focused on preventing loss through updating infrastructure and investments in expanding supplies through construction of reservoirs and desalination plants. This has not been accompanied by a major push to limit demand through reduction of consumption by end users. The narrow measures taken to limit demand have been based on pricing disincentives and the installation of meters in new builds following legislative requirements.

This has resulted in a situation where drought management systems are strong at the corporate and public policy levels but bear little relevance at the local level. This is reflected in the findings of the Drought LAC index. Water Reliant businesses (often the most intensive users of water) are metered but completely exempt from hose pipe bans or any drought measures during times of water scarcity. This has led to little incentive for private and corporate water users to reduce water consumption or evolve drought resistant practices. Interviews revealed an awareness of increasing drought risk and changing climate conditions among local practitioners, but little action at an organizational level to deal with such potential hazards. There was also minimal collaboration between water companies and end users (whether household consumers or big businesses) to implement risk reduction or adaptation practices. To some degree this was attributed to the closed and engineering focused management attitudes of water companies, which were seen as unapproachable or uninterested in local level collaborations.

In addition, past experiences of drought in other regions revealed little consideration of social and health risks posed by drought orders, which were not considered in drought planning processes since the brunt of management efforts focused on drought prevention rather than drought management. This was due to an assumption on the part of water companies that moving beyond a hose pipe ban would be avoided through a manipulation of supplies rather than managing demand. This aversion to imposing constraints on public demand can be seen as an indirect consequence of the privatization of water provision.

Overall the index, both at the risk management and local level, proved very useful in providing an assessment of latent vulnerability to heatwaves and drought in London. It provided a picture of management robustness not only for the purpose of the study but also for study participants and practitioners to apply to their own context. The application of the index was also a simultaneous validation exercise since respondents participated in its development and ultimately utilized it to reflect on and improve their own practices.

Although specific physical, geographical, and social determinants affect the relevance of some indicators over others, the index is a useful tool for a contextualized assessment of what is strong, what is weak, and what is needed for a robust risk management system that alleviates vulnerability to drought and heatwave hazards. The combination of a structured index with open-ended interviews allowed for a comprehensive qualitative assessment of risk management processes, circumstances and needs, while anchoring discussion in the index structure to provide comparative and measurable outcomes and data. It offers an interesting methodology for use in vulnerability assessments for the MOVE project.

6.7 CONCLUSIONS AND RECOMMENDATIONS

The RMI and LAC indexes in conjunction with semistructured interviews with risk management practitioners, vulnerable groups, and experts have proved to be a quick and effective method of assessing vulnerability in a particular system through an examination of the strengths and weaknesses of its risk management structure. Although this methodology does not provide a quantitative measure of existing vulnerability, it provides qualitative and actionable assessments that allow for policy and practical improvements to strengthen the overall risk management and disaster response mechanisms. Through its application to heatwaves and drought hazards in London, the following observations and recommendations are made with regards to risk management and vulnerability assessment:

- Vulnerability to climate change hazards is a dynamic function of both physical as well as social factors. Both types of indicators need to be taken into account when measuring vulnerability.
- Disaster risk management directly impacts vulnerability and requires multi-agency coordination from organizations involved not only in disaster response but also risk reduction (such as social services).
- Disaster management plans need built-in mechanisms to ensure that they move beyond their role as risk management regimes to best-practice guidelines for local actors.
- Climate change hazards need greater prioritization and given further policy and financial support by governments in order to decrease vulnerability and risk.

REFERENCES

Abrahamson, V., Raine, R., 2009. Health and social care responses to the department of health heatwave plan. Journal of Public Health, 1–12.

Abrahamson, V., Wolf, J., Lorenzoni, I., Fenn, B., Kovats, S., Wilkinson, P., Adger, N., Raine, R., 2008. Perceptions of heatwave risks to health: interview-based study of older people in London and Norwich, UK. Journal of Public Health, 1–8.

Basu, R., Samet, J.M., 2002. Relation between elevated ambient temperature and mortality: a review of the epidemiologic evidence. Epidemiologic Reviews 24 (2), 190–202.

Bell, M., O'Neill, M., Ranjit, N., Borja-Aburto, V., Cifuentes, L., Gouveia, N., 2008. Vulnerability to heat-related mortality in Latin America: a case-crossover study in São Paulo, Brazil, Santiago, Chile and Mexico City, Mexico. International Journal of Epidemiology 37, 796–804.

Bond, N., Lake, P., Arthington, A., 2008. The impact of Drought on freshwater systems: an Australian perspective. Hydrobiologia 600, 3–16.

Cardona, O.D., 2005. Indicators of Disaster Risk and Risk Management- Program for Latin America and the Caribbean: Summary Report. Inter American Development Bank, Department of Sustainable Development, Washington.

Conti, S., Meli, P., Minelli, G., Solimini, R., Toccaceli, V., Vichi, M., Perini, L., 2005. Epidemiologic study of mortality during the summer 2003 heat wave in Italy. Environmental Research 98 (3), 390–399.

Environment Agency, 2007. Areas of Water Stress: Final Classification. Environment Agency, UK. http://publications.environment-agency.gov.uk/pdf/GEHO1207BNOC-e-e.pdf.

Faunt, J.D., Wilkinson, T.J., Aplin, P., Henschke, P., Webb, M., Penhall, R.K., 1995. The effete in the heat: heat-related hospital presentations during a ten-day heat wave. Australian and New Zealand Journal of Medicine 25, 117–120.

Fitzhugh, T., Richter, B., 2004. Quenching urban thirst: growing cities and their impact on fresh water ecosystems. Bioscience 54 (8), 741–754.

GLA, 2006. London's Urban Heat Island: A Summary for Decision Makers. Greater London Authority, London. http://legacy.london.gov.uk/mayor/environment/climate-change/docs/UHI_summary_report.rtf.

GLA, 2007. Water Matters. Greater London Authority, London.

Hajat, S., Armstrong, B., Baccini, M., Biggeri, A., Bisanti, L., Russo, A., Kosatsky, T., 2006. Impact of high temperatures on mortality: is there an added heat wave effect? Epidemiology 17 (6), 632–638.

Hajat, S., Kovats, R.S., Lachowycz, K., 2007. Heat-related and cold-related deaths in England and Wales: who is at risk? Occupational and Environmental Medicine 64 (2), 93–100.

Heathcote, R.L., 1991. Managing the droughts? Perception of resource management in the face of the drought hazard in Australia. In: Vegetation and Climate Interactions in Semi-arid Regions. Springer, Netherlands, pp. 219–230.

Hewitt, K., 1997. Regions at Risk. A Geographical Introduction to Disasters. Addison Wesley Longman Limited, England.

McGregor, G., Pelling, M., Wolf, T., Gosling, S., 2007. Social Impact of Heatwaves. Science Report SC 20061/SR6. Environment Agency, UK.

Meehl, G., Tebaldi, C., 2004. More intense, more frequent, and longer lasting heat waves in the 21st century. Science 305, 994–997.

Mpelasoka, F., Hennessey, K., Jones, R., Bates, B., 2008. Comparision of suitable drought indices for climate change impact assessments over Australia towards resource management. International Journal of Climatology 28, 1283–1292.

Newton, S., 2005. Design of Environmental Conditions in Hospital Wards. Case Study: Four Wards in Whittington Hospital. Master of Science, University College London, London.

Stafoggia, M., et al., 2006. Vulnerability to heat-related mortality: a multicity population-based case-crossover analysis. Epidemiology 17, 315–323.

Wolf, T., 2009 Assessing vulnerability to heat stress in urban areas (unpublished Ph.D. thesis), King's College London.

Worfolk, J.B., 2000. Heat waves: their impact on the health of elders. Geriatric Nursing 21, 70–77.

Chapter 7

Comprehensive Vulnerability Assessment of Forest Fires and Coastal Erosion: Evidences from Case-Study Analysis in Portugal

Dr Fantina Tedim*, Dr Manuel Garcin†, Dr Charlotte Vinchon†, Dr Salete Carvalho*, Dr Nicolas Desramaut‡, Dr Jeremy Rohmer§

*Department of Geography, Faculty of Arts, University of Porto, Porto, Portugal, †BRGM, Risks and Prevention Division, Coastal Risks and Climate Change Unit, Orléans, France, ‡BRGM, Risks and Prevention Division, Ground Instabilities and Erosion Risk Unit, Orléans, France, §BRGM, Risks and Prevention Division, Risks of underground Storages and Exploitations, Orléans, France

Chapter Outline

Assessment of Vulnerability to Natural Hazards. http://dx.doi.org/10.1016/B978-0-12-410528-7.00007-2

7.1 VALIDATION OF THE VULNERABILITY FRAMEWORK: METHODOLOGICAL QUESTIONS

The validation of MOVE's framework on the North-western of Portugal, focusing on coastal erosion and forest fires, raised three methodological questions.

Firstly at which scale should it be implemented? The selection of "local scale" to support vulnerability analysis was mainly justified by the increasing responsibilities of municipalities in the assessment and management of natural risks. The test site for coastal erosion was the municipality of Esposende (Figure 7.1). This municipality has a surface of 95 km^2 and 34,254 permanent inhabitants (INE, 2012). However, this area has numerous second houses and touristic infrastructures whereby the population can hugely increase during weekends and during the summer. Its coastline with 16 km length, fringed by the Atlantic Ocean, presents one of the highest erosion rates in Portugal. To protect assets, too close to the shoreline, coastal defense structures were built (e.g., rip raps, groynes).

Six municipalities from a mountain area were selected as forest fires test case (Arcos de Valdevez, Melgaço, Montalegre, Ponte da Barca, Terras de Bouro, and Vieira do Minho). In 2011, 74,912 inhabitants lived on the 2,200 km^2 covered by these municipalities (INE, 2012). These municipalities are characterized by low population density and small urban centers. Depopulation and ageing have been increasing in most of the 167 parishes. Wildlands account for 80% of total surface while agriculture accounts for 14% of total surface (IGEO, 2007). Some of this area belongs to the Peneda-Gerês National Park, the only national park existing in the country.

The second question was related to the purpose of vulnerability assessment. In the coastal erosion test area, the purpose was to identify the vulnerability of main functions of parishes (freguesias) for three different scenarios. Concerning forest fires, the aim was to produce a score/rating of a community's vulnerability, considering all components and dimensions in order to produce a vulnerability map. This map when integrated with hazard map would produce a true risk map. The existing "risk" map elaborated at municipal scale focusing only on hazard assessment has been influencing landscape management planning and regulation with high implications in households' daily life.

The final question was to what extent should the stakeholders participate in the framework validation? Although conscientious of the distance between the citizens and the power institutions as well as the uncertainty avoidance that characterize Portuguese culture (see http://geert-hofstede.com/portugal.html) an active participation of stakeholders in all research process was regarded as adequate. For each test case, the stakeholders were identified and considering their different responsibilities and competences in risk management they were aggregated in three groups: *policy level* responsible for the definition of regulation and technical guidelines (e.g., ICNF—Institute for Nature Conservation and ANPC—Civil Protection National Authority); *operational level* responsible for the implementation of risk management policies (e.g., Municipal Civil

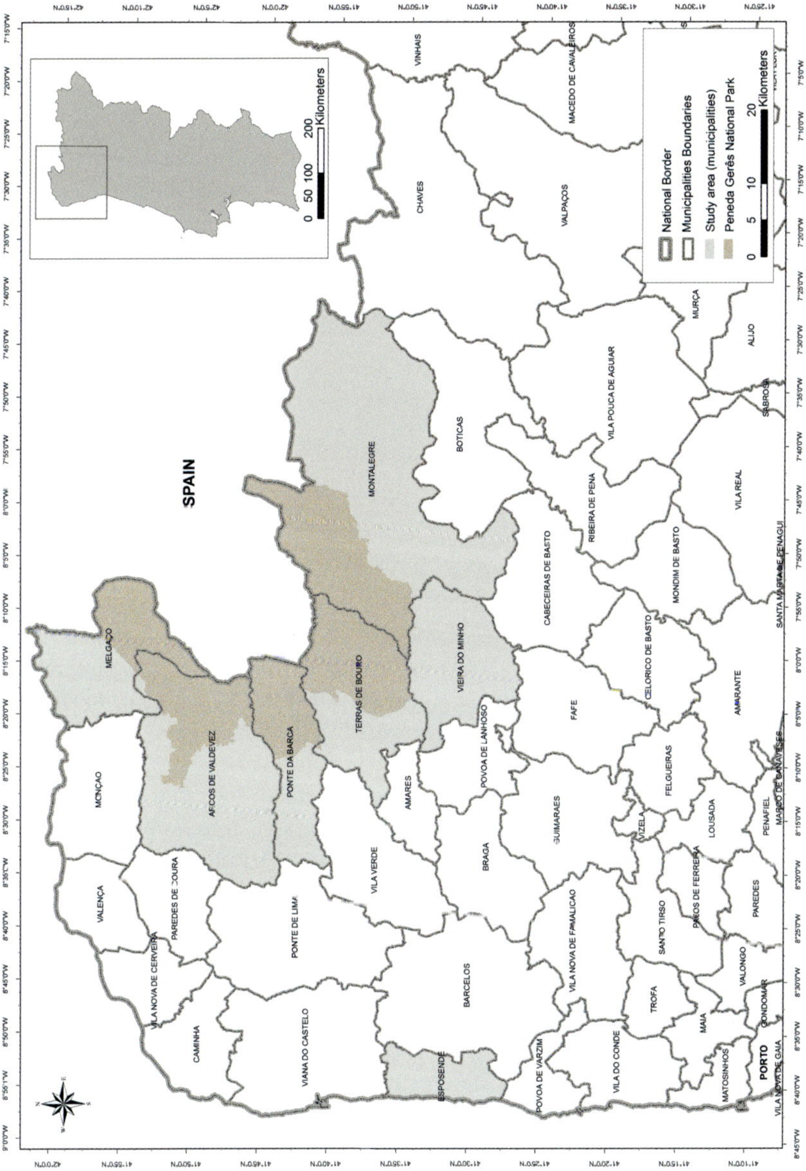

FIGURE 7.1 Location of the two test sites of the NW Portugal case study.

Protection Service of each municipality, North Littoral Natural Park, Peneda-Gerês National Park, Fire brigades); *civil society* which are the target of the risk management policies (e.g., residents, landowners, farmers).

The stakeholder's participation started with the selection of vulnerability indicators. For this task, stakeholders from the operational level were involved because they are closer to the reality and have a good awareness of local problems (e.g., through contact with the citizens) and the virtues and constraints of legal prescriptions from an empirical basis. In the first workshop, the MOVE's framework was introduced. This included explaining the concepts of vulnerability and resilience which they were not familiar with; as well as the indicators list which was developed using deductive, inductive, and normative procedures as suggested by Hinkel (2011). This presentation generated a significant interest among the stakeholders that showed availability to participate in more workshops and to be actively involved in the process of building a vulnerability map for their municipalities.

7.2 HAZARDS CHARACTERIZATION

7.2.1 Coastal Erosion

Erosion rate is highly variable along the Esposende municipality coast in relation with the morphology, the geological and hydrological context, the lithology of each coastal stretches, (Loureiro, 2006) as well as with the coast defense structures. The main observed effects of the erosion are a retreat of coastline, changes in beaches profiles, destruction of sandy beaches and dunes, land losses, damages and destruction of buildings and infrastructures, and indirectly an increase of the marine submersion hazard. The forcing factors of local erosion are the littoral drift and waves (erosive crisis during storms), but the main cause of the long-term erosion is due to a regional negative sedimentary budget and a weak physical resilience due to lack of sediment stock in the shoreface.

The sandy beaches are generally affected by seasonal phases of erosion during winter (due to wave action during storms) and aggradation during summer. In Esposende coast, the lack of sediment stock in the shoreface does not allow the reconstruction of the profile range during summer. The existence of littoral drift can lead to a long shore transport of sediment along the coast. In some coastal stretches, the sedimentary budget (at the pluriannual scale) is globally positive inducing aggradation, on other coastal stretches; the sedimentary budget is null inducing stability. In the case of negative sedimentary budget which is the most common, the coast is in erosion.

Thus, at the annual scale, erosion can be described as an oscillatory process around an average position of the coastline, with erosional events which can be "crisis erosional event" followed by a relaxation phase with sedimentary aggradation which tends to restore position of the coastline which prevails before erosion crisis. The space of oscillations can be presented as an uncertainty strip around a medium coastline position. This uncertainty strip can represent at a short time scale the area exposed to the erosion hazard (Crisis event: short-term erosional event with coastline retreat during a storm) (Figure 7.2).

At longer time scale (pluriannual) and constant hydrodynamic conditions, the spatial evolution of the coastline is dependent on the sedimentary budget of the concerned coastline whatever the processes acting on this budget (i.e., littoral drift; trapping of sediment by defense works: groynes, breakwaters, dykes; deficit of sedimentary transport by river linked to river damming). Thus, "long term" coastal erosion can be characterized by a pluriannual average rate (i.e., long-term erosion: coastline erosion rate (average) based on multidecadal evolution).

Coastal erosion, in itself, is a reversible process at some timescales; the accretion phase can occur after erosion phase and lead to a coastline recovery. The physical resilience of a coastal strip is linked to the stock of sand and to the availability of this sand offshore and backshore for same conditions. Thus, the role of the sand dunes is of a primary importance to the resilience capacity of the coast.

However, even if coastal erosion is reversible, the effect of coastal erosion on assets is irreversible. The buildings, roads, other infrastructures, natural habitats, etc. located on eroded lands are irremediably destroyed, lands and soils are definitely lost.

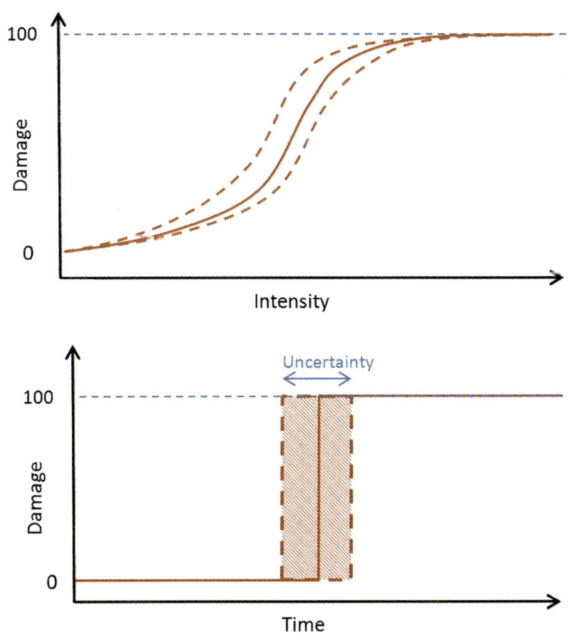

FIGURE 7.2 Shape of the fragility curve for numerous natural hazards (top) and specific fragility curve for coastal erosion (bottom).

Hence, in this approach, the fragility of assets to coastal erosion is assumed as Boolean an asset which is affected by the coastal erosion is definitely destroyed and damaged and unusable. Thus, we can consider that the fragility of assets (e.g., buildings, networks, roads) is a Boolean function with only two damage states: *Not Affected* or *Affected*. For this specific hazard, the vulnerability is directly linked to exposure to the hazard.

Therefore, the return period is a concept not adapted to the coastal erosion hazard. Instead, for a specific location we propose the concept and use of the *time expectancy*. For the application to the Esposende site, we chose a 30 years time expectancy consistent to the duration of the planning landscape projects. Noteworthy, in the frame of the MOVE project we have not taken into account the climate change (e.g., sea level rise, storm surge) which is an important component potentially acting on erosion processes.

In order to locate the location of the 30 years time expectancy line, we have determined the erosion rates for each homogeneous coastal segment using data from Loureiro (2006). The analysis of the cross-shore profiles shows that the rate at each point can vary in time. In Esposende, the duration of recorded data concerning the evolution of the coastline due to erosion is too short to determine an average accurate and reliable rate.

Hence, to compensate the lack of observations which would have given a good appreciation of the velocities of the coastline retreat, we adopted a scenario-based approach in order to estimate the range of erosion rates of the coastline. The scenario approach rely on the combination of the available data on recent (1994-2004) coastline evolutions (Loureiro, 2006), the observations carried out on field during the MOVE project, and the knowledge of local experts about coastal processes. Thus three hypotheses of Coastal Erosion Rate (CER) are determined for the next 30 years. The likelihood of occurrence of each of them is qualified (i.e., associated with a qualitative uncertainty level): (1) a *Very likely* coastal erosion rate, (2) a *Likely* coastal erosion rate, (3) a *Possible* coastal erosion rate.

Using GIS, we projected the areas which will be affected by erosion in the next 30 years for each of the three categories of uncertainties: *Very Likely*, *Likely*, *Possible* using the associated CER (Figure 7.3).

7.2.2 Forest Fires

Forest fires are the main threat to productivity and sustainability of Portuguese forests. Most of the fire outbreaks are anthropogenic. Anthropogenic influence is manifested by starting new fires, changing fire behavior (e.g., managing fuels, forest flammability), suppressing fires (e.g., firefighting resources), and people preparedness to face fire outbreak. The increase of fuel accumulation and ecosystems combustibility are a result of socioeconomic changes (e.g., depopulation, ageing, land abandonment, changes in the use of forest products, expansion of high flammable species

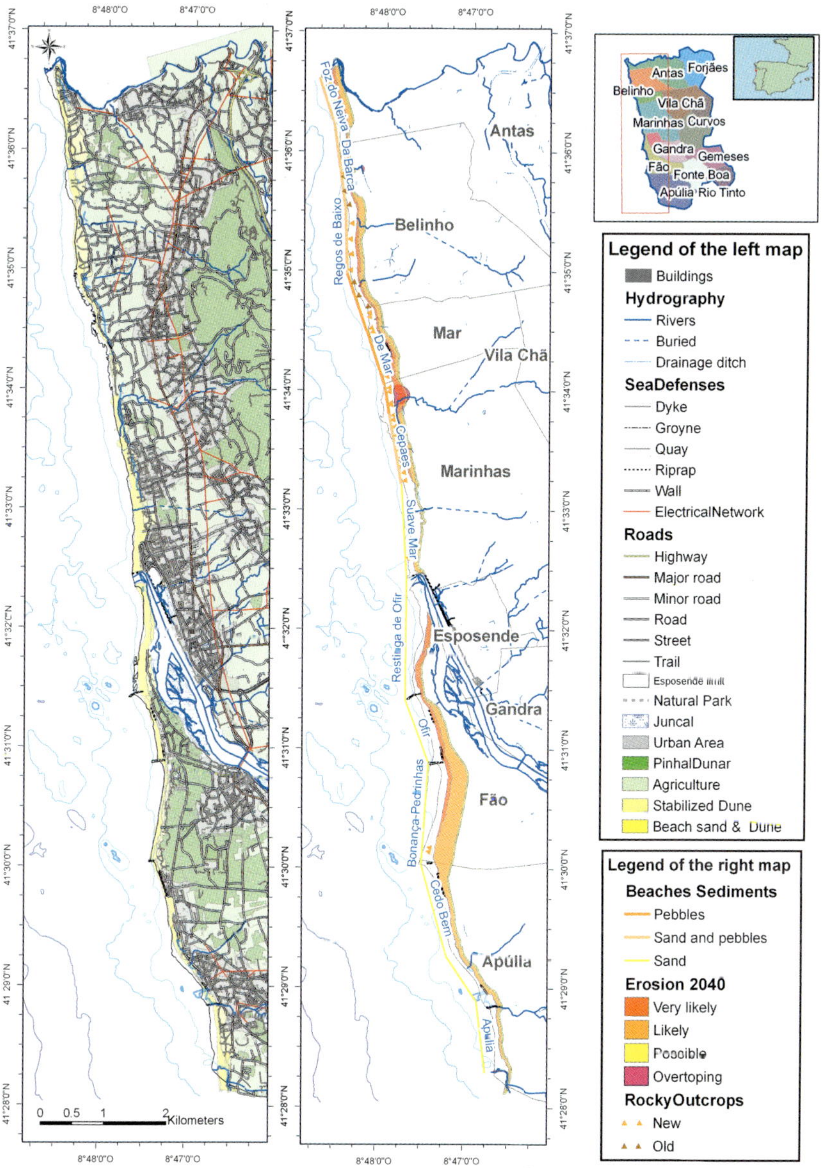

FIGURE 7.3 Land use, assets, and coastal erosion hazard for the Esposende municipality.

like pinus and eucalyptus). The expansion of wildland–urban interface (WUI) is putting more people at risk and increasing potential damages. It is a consequence of the expansion of urbanization to wildland but also of the abandonment of agriculture allowing wildland to come closer to the settlements.

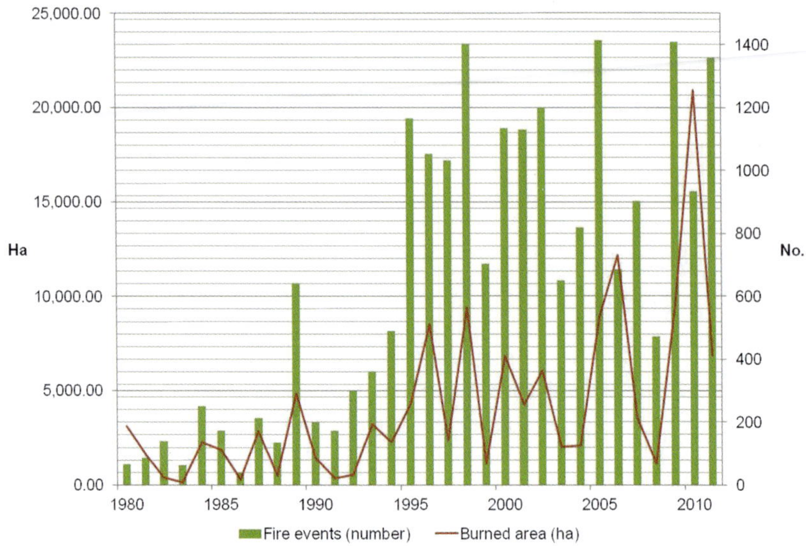

FIGURE 7.4 Forest fire occurrence and burned area between 1980 and 2011. *Source: Data from ICNF (2012).*

In the Portuguese test case between 1980 and 2011, 20,609 fires were registered, burning 135,934 ha (ICNF, 2012). A trend in the increase of fire occurrence and burned area is evident (Figure 7.4).

The fire impacts are generally assumed, by the fire agencies, as related with fire size. However, burned area is neither well correlated with fire ecological severity (Lutz et al., 2011) nor with the social impacts (Gill and Moore, 1998) and just because a fire is large it does not necessarily mean that it is a disaster (Keane et al., 2008; Tedim et al., 2013). Most fires (almost 99%) have a burned area of less than 100 ha (46% of total area). The fires larger than 1,000 ha represent less than 1% of the total of occurrences and 19% of the burned area.

Most of the surface of the six municipalities was affected by fires in last 35 years (Figure 7.5). Only in Terras de Bouro the burned area is inferior to the municipality surface (80% of the municipality surface). On the other municipalities, the percentage of burned area is much higher (e.g., 162% for Vieira do Minho), meaning that some areas burned more than once. Just 24% of the surface was affect by a fire only once and about 15% burned more than five times. In some places, the periodicity of fire occurrence was 2–3 years (Figure 7.5) which prevents any return in forest investment.

7.3 VULNERABILITY ASSESSMENT METHODS

7.3.1 Coastal Erosion

The time expectancy approach is applicable to the exposed lands, to the ecosystems, and also to the human assets (e.g., buildings, roads, infrastructure, and

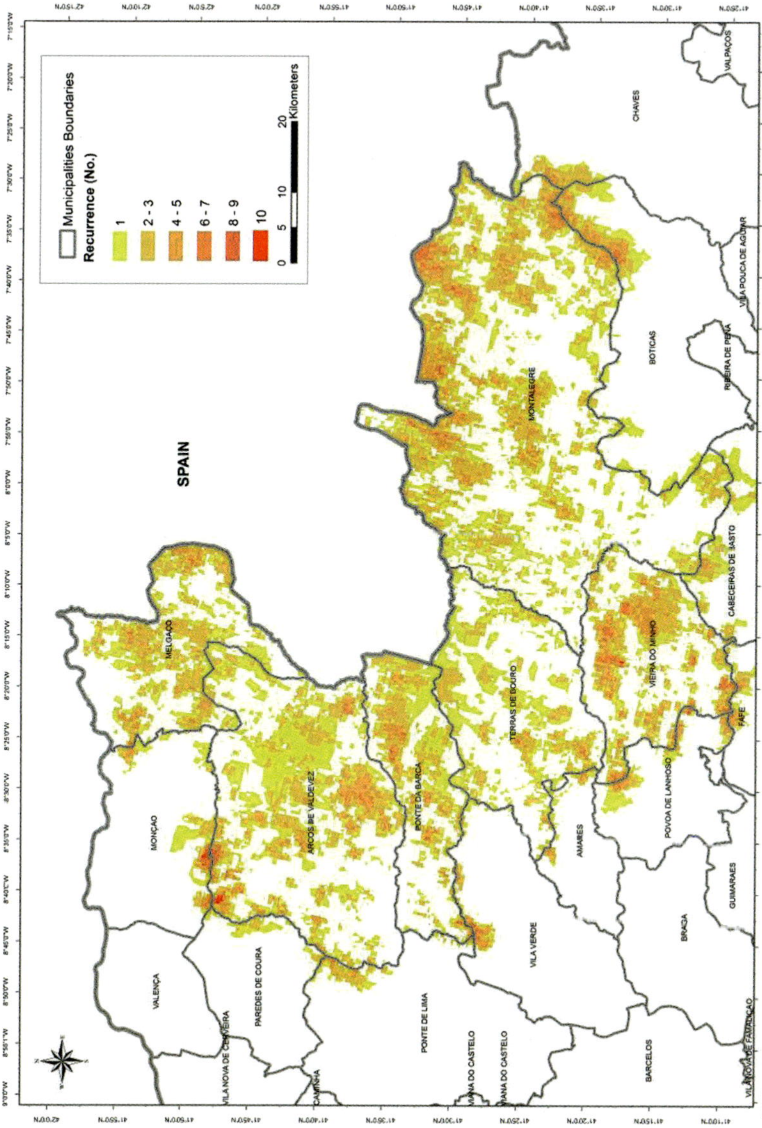

FIGURE 7.5 Forest fire recurrence between 1980 and 2011. *Data from ICNF (2012).*

population). In order to evaluate the exposure of assets to coastal erosion hazard, we have used the three sets of zoning giving us the assets exposed at erosion hazard in 2040 associated with their uncertainty level (Very likely, Likely, Possible). The exposure of different types of assets has been evaluated in order to quantify the number of affected assets for each parish (Figure 7.6). The results can be presented on maps or on graphs and sorted using their uncertainty levels.

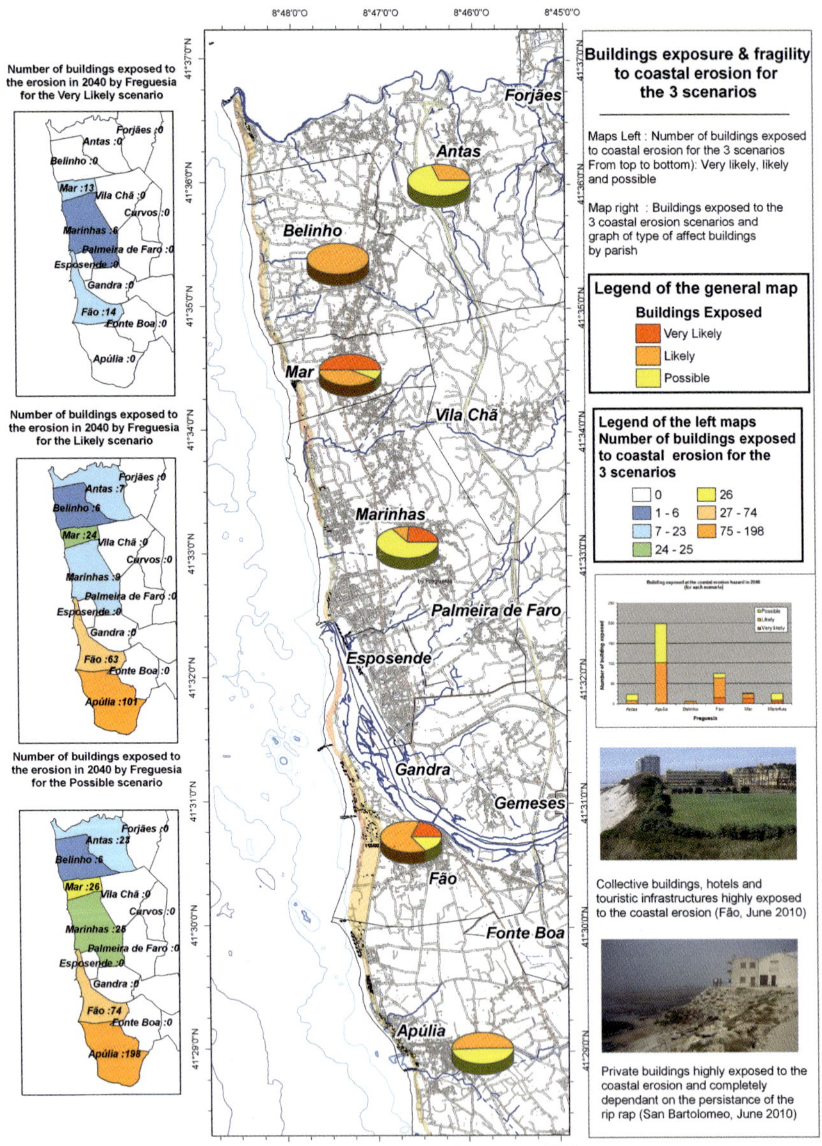

FIGURE 7.6 Number of buildings exposed to coastal erosion in 2040 for the three scenarios.

It is then possible to compare the exposure of coastal erosion for each type of assets and for each parish.

Instead of using single indicator with a numerical value attached to each individual asset, it is possible to have a vulnerability assessment using the surface of specific land use potentially eroded and destroyed. Each land use class is in fact an aggregating indicator at which the dominant function of the land is affected. The erosion of a surface of a specific land use has a direct effect on the linked function within the community and the society. For example, ecological vulnerability for each parish can be expressed as the percentage of "natural" land (active dune, vegetated dune, and forest) potentially destroyed by erosion in relation with total "natural" surface for each parish. The vulnerability of agriculture can be expressed as a percentage of potentially destroyed agricultural land in relation with the total surface of agricultural land for each parish, and urban infrastructure vulnerability can be expressed as the percentage of urban areas potentially destroyed by the total of urban areas.

7.3.2 Forest Fires

In the development of forest fires vulnerability assessment, population, structures, activities, and ecosystems were considered as exposed elements. For each of them, explanatory variables of exposure, fragility, and lack of resilience were identified. Afterward for each variable, indicators were identified. A broad list of indicators was created based on their interest in explaining vulnerability without any restrictions considering data availability.

While indicators reduce society and environmental systems complexity, they are useful to communicate complex issues from science to policy makers and the general public. It was our purpose to develop indicators that could be, at the same time, scientifically sound, easy to interpret, understandable, and policy relevant. Although we started looking for indicators useful to assess the forest fire vulnerability in Portugal, we soon realized that concerning the goal and the spatial scope of the MOVE project it could be useful to consider different geographical and cultural contexts and propose indicators that could be applied in other European countries and even around the world to extend the impacts of the framework.

The proposed list of variables and indicators was developed with the purpose of being an *open frame* to: (1) accommodate new development produced by scientific community or empirical evidences; (2) be adapted to different geographical and cultural contexts; (3) support the model application from the household to national scale.

Through four workshops, the stakeholders validated the indicators selecting the ones most suitable to the Portuguese context. Other issues were discussed like the nonexistence of appropriate secondary data set to develop some indicators, the possibility to develop primary information, the best way to measure and weight each indicator to build a vulnerability map using GIS tools. The last workshop was open to all types of stakeholders. Two main subjects became the focus of discussion: (1) the operationalization of the framework, the main concern being the indicators rating process; (2) a brainstorming session focusing on the various barriers and opportunities associated with the applicability of this model of vulnerability assessment and its contribution to enhance risk management practice in Portugal (Paton and Tedim, 2012).

For the operationalization of the indicators list, the municipal scale was our basis of analysis. Each indicator, adequate for Portuguese context, for which information was available, was mapped for the test area in order to evaluate its soundness and the best way to map it. The indicators developed were mainly quantitative but in the mapping process most of the variables were ordinal. Spatial analysis using GIS was the main tool used to aggregate the data and map the vulnerability indicators. Although the final purpose is to produce a global score of vulnerability useful to produce a true risk map, each indicator proposed can be considered on its own because it has its own meaning.

In order to appreciate the validity of the indicators selected in different cultural contexts, the list of indicators was sent to five international experts on forest fires from Bulgaria, France, Greece, Chile, and USA and four national experts, to be evaluated by them using a Likert scale (discussed in Section 7.6).

7.4 INDICATORS DESCRIPTION

7.4.1 Coastal Erosion

7.4.1.1 Index of Coastal Erosion Exposure and Fragility

We define the coastal erosion vulnerability index as the sum of the alteration of three functions of the community: *agriculture*, *urban* (e.g., housing, facilities, energy, health, education), and *ecological*.

The alteration of the functions is evaluated using the percentage of the lands affected by erosion relatively to the total surface of area affected to this function in the parish (normalization). This is realized for the three scenarios of erosion of land in the next 30 years (Very likely, Likely, Possible).

Two parishes can have the same vulnerability index but in closer detail, they can be completely different (e.g., agricultural function for the first and urban function for the second).

The graphical presentation of radar (Figure 7.7) can be used to capture an overview of the characteristics of the vulnerability of each parish. In the radar, the three axes correspond to one family of vulnerability (i.e., *ecological*, *agriculture*, and *urban*). The radar shape of each parish indicates the main characteristics of the vulnerability (dominance of one of the factor for example), the total surface indicating the "intensity" of the vulnerability. One of the more interesting aspect of this approach is to identify the variability both quantitatively and qualitatively of the vulnerability.

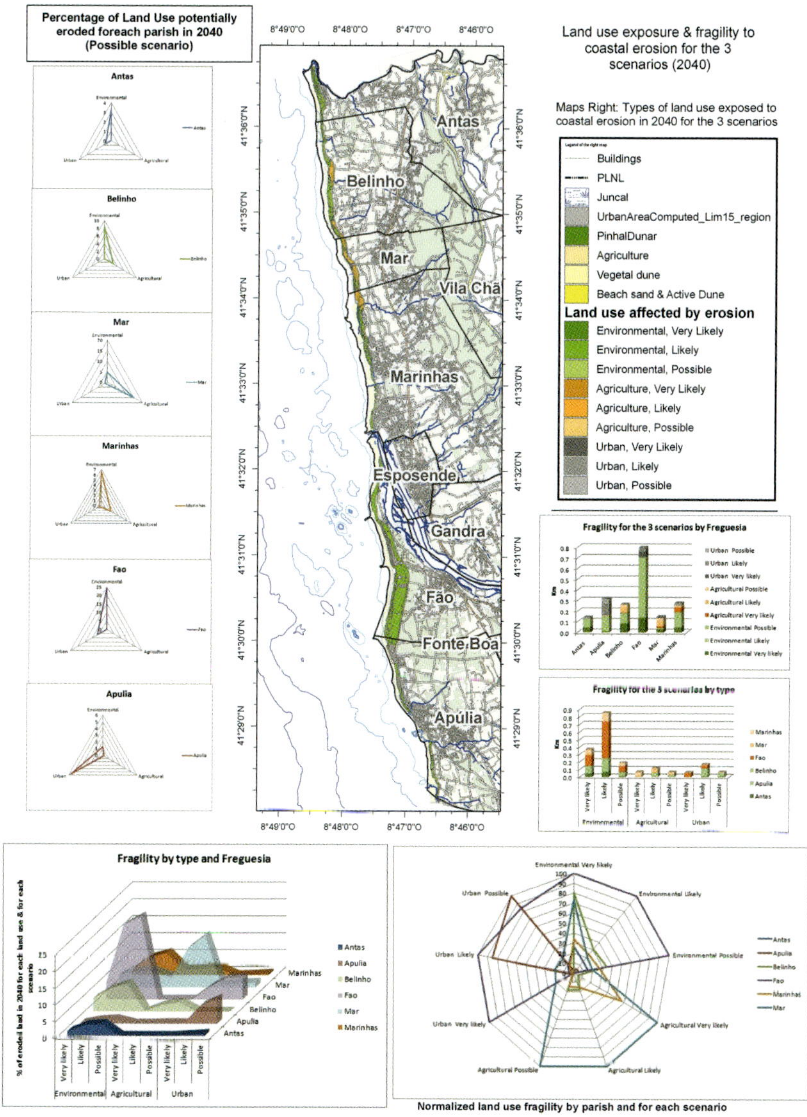

FIGURE 7.7 Land use exposure and fragility of parishes to coastal erosion.

Thus, the Fão parish seems to be more vulnerable to the coastal erosion, being ecological and urban functions the most affected. On other hand, Antas seems to be the less vulnerable. In this parish, the ecological assets are particularly vulnerable.

Analysis of the radars and histograms (Figure 7.7) shows more variability of the parish vulnerability. Each parish has its own dominant vulnerability: *urban* for Apulia, *agriculture* for Mar, *ecological* for Belinho and Marinhas.

In addition, some of the parishes show a bimodal vulnerability *urban-ecological* and *ecological-agriculture*.

7.4.1.2 Indicators of Coastal Erosion Resilience

Using the MOVE framework, we have constructed a table of the resilience indicators (Table 7.1).

Five main types of resilience have been identified respectively: physical, environmental, economic, social, and institutional. The resilience capacity can be decomposed into *capacity to anticipate*, *to cope*, and *to recover*.

The resilience to erosion at the site of Esposende has been analyzed using these various types. We note that there are many indicators of resilience for the ability to recover while few indicators exist about the capacity to anticipate.

7.4.1.2.1 Physical Resilience Indicator

The physical resilience to coastal erosion can be expressed as the volume of sand present on the coastline. Ideally, this indicator must integrate the sedimentary stocks offshore (fore beach), on the beach, and backshore. Unfortunately, data about the volume of sand and gravel of the foreshore are inexistent. But, the field observations and of the ortho-photography are showing that the sandy stock is very poor; the sand of some beaches of the Esposende municipality have been recently replaced by pebbles. On the foreshore, numerous rocky outcrops are visible indicating that the sand has been eroded. When the sands are present, they seem to be residual.

The sedimentary stock of the backshore plays an important role as a sediment supplier and is mobilized by dynamic processes responsible for erosion, transport, and accretion along the coast. The higher the sedimentary stock backshore, the higher the coast is resilient to erosion. The sedimentary stock affects both the capacity to cope and the capacity to recover.

Thus, we are considering that the presence of the dunes (both active and vegetated) is an indicator of physical resilience. We can consider that the average dune wideness is a general indicator of the actual resilience of a coast segment. Thus, for each parish we have computed the average dune width (Figure 7.8)

Figure 7.8 shows that the widest dunes are present in Belhinho, Antas, and Marinhas while the narrowest are in Mar. This suggests that the Mar parish is physically less resilient to coastal erosion while Antas and Belinho are more physically resilient. The dunes width for each parish has been normalized and the relative resilience of the parish can then be compared (Figure 7.9). The average width of the dunes for each scenario and for each parish was also computed. This illustrates the evolution of the physical resilience for the three erosion scenarios. One can notice that the evolution of the physical resilience is not identical for each parish (Figure 7.9 and Figure 7.10).

In some parishes, the actual physical resilience will be relatively slightly changed while some parishes show an important decrease of the resilience indicator in the future. The influence of the scenario used can have very different effects on the future resilience indicator. For example, the impact of the scenario choice is relatively low for Mar, values obtained largely lower than today but are

TABLE 7.1 Resilience Indicators for Coastal Erosion

Resilience Type		Resilience Capacity: Capacity to Anticipate	Capacity to Cope	Capacity to Recover
Physical resilience			Sedimentary stock (foreshore, beach, backshore)	Sedimentary stock (foreshore, beach, backshore)
			Coastline geology & geomorphology	Coastline geology & geomorphology
Environmental			Ecosystems adaptability/ endemism	Ecosystems adaptability/ endemism
Economy	Economic status	Availability of coastal protection works funds, GDP		GDP, resources, availability of coastal protection works funds
	Empowerment (autonomy)			Economic autonomy of citizen versus economic dependence to community
				Economic autonomy of community versus economic dependence to state
	Companies		Financial resource of companies	Financial resource of companies
				Insurance rates
				Insurances and state compensations
Social			Population age	Population age
		Education	Education	Education
				Socialization/Sense of community
Institutional		Landscape planning regulation		

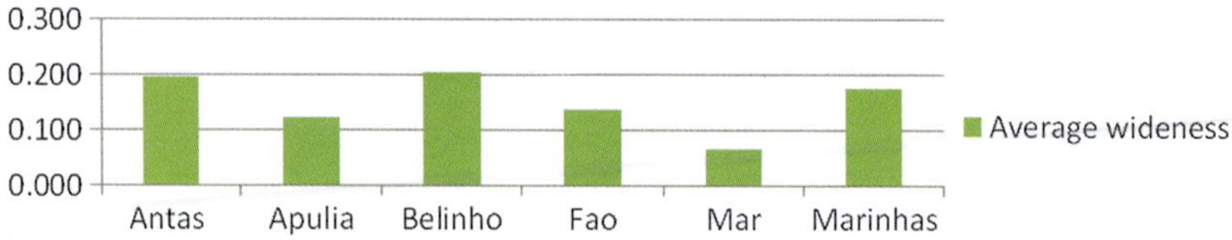

FIGURE 7.8 Average width of dunes by parish (vertical unit in km).

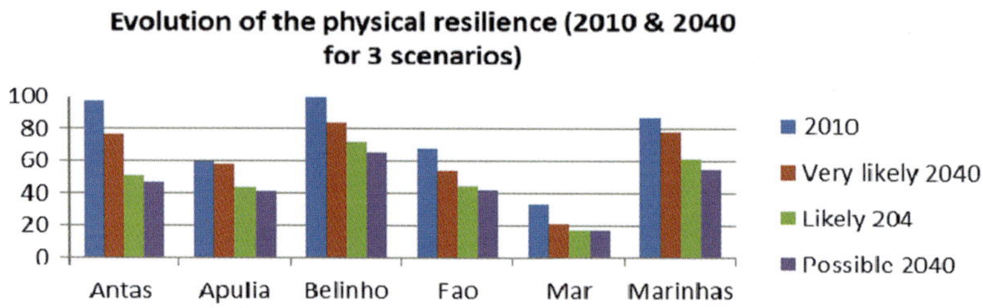

FIGURE 7.9 Physical resilience indicator to coastal erosion.

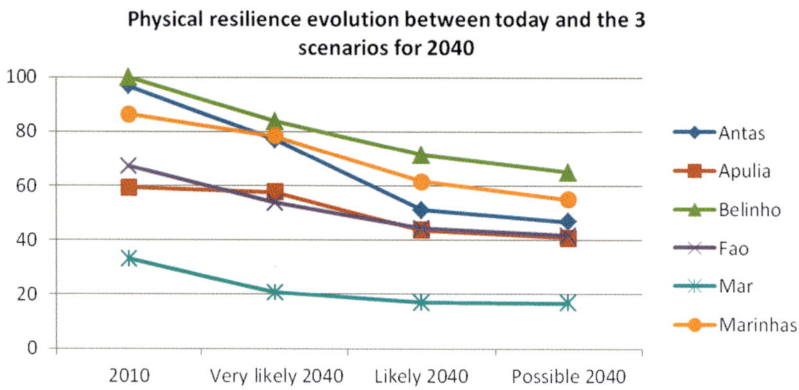

FIGURE 7.10 Evolution of the physical resilience to coastal erosion for today and in 2040 for the three scenarios.

relatively close one to each other's. For Apulia, the value of the indicator found in the case of the very likely scenario gives a value relatively closed to those of today while the values obtained for the likely and possible scenarios are largely lower than today. Other example with Antas have one of the stronger resilience indicator for today but shows a very important decrease of value for the very likely scenario and even more for the likely scenario.

The analysis of the physical resilience indicator shows the variability of the value of the indicator between the parishes. Additionally, it demonstrates that the physical resilience is not static and can change in the future. This future evolution will not be identical for all parishes and must be analyzed specifically.

TABLE 7.2 Exposure Indicators

Variables	Indicators
Proximity to the forest and extensive grazing areas	Wildland–urban interface characteristics (HD)[1] Distance of each settlement to the forest (HD)[1] Distance to extensive grazing areas (HD)[1]
Exposed population and buildings	Population density (HI)[1] Number of households (HI) Number of buildings(HI)

HD—Hazard dependent indicator; HI—Hazard independent indicator.
[1]*Notes: The underlined indicators are the ones used in the elaboration of a forest fire vulnerability map at the municipal scale in Portuguese test case.*

7.4.2 Forest Fires

The primary ecological and social effects of forest fires are physical: human structures and vegetation burned and people injured and/or dead (Bovio et al., 2006). This is a consequence of exposure but also of susceptibility/fragility and lack of resilience.

7.4.2.1 Exposure Indicators

Table 7.2 presents the exposure variables and indicators, the ones validated in the test case are underlined.

Although there is no single definition of WUI, normally it includes three components: human presence, wildland vegetation, and a distance that represents the potential for effects (Stewart et al., 2007). The coexistence of buildings and dense vegetation increases the probability of fire ignition favoring fire spread. The *WUI characteristics* as well as the *distance of each settlement to the forest* create different levels of exposure. The *distance to extensive grazing areas*, where pasture renovation using fire is a common practice, decreases the exposure to forest fires.

The higher the amount of population (e.g., *population density*) and buildings (e.g., *number of buildings* or *households*) exposed which can potentially be affected by fires, the higher the potential damages. A high population density equals to higher evacuation difficulties.

The exposure evaluation contributes to strategic fuel management, planning emergency response, and identifies the needs in prevention and preparedness.

7.4.2.2 Susceptibility/Fragility Indicators

7.4.2.2.1 Physical

Physical susceptibility/fragility (Table 7.3) is mainly explained by *structural forest defensibility* and *buildings susceptibility*. The former can be evaluated

TABLE 7.3 Susceptibility/Fragility Indicators

Dimension	Variables	Indicators
Physical	Structural forest defensibility	Forest areas slope (HI)[1] Species flammability (HD)[1] Forest accesses and fuel breaks (HD)[1]
	Buildings susceptibility	Buildings ignition potential (HD) Housing security (HI)
Social	Population characteristics	Elderly (people over age 75) (HI)[1] Functional age (HI) People disability (HI) One person household (HI)
	Deprivation and social exclusion	Deprivation index (HI)
	Integration in communities	Temporary residents and holidaymakers (HI) Ethnic minorities' integration (HI) Migrants in country for less than 2 years (HI)
	Chronic health conditions	Cardiovascular and respiratory diseases (HI) People with mental disorders (HI)
	Absence or lack of motivation/opportunity	Dwellings used as first residence (HI)[1] Absentee homeowners (HI) Absentee landowners (HI) Renter houses (HI)
	Community fragmentation	Existence of conflicts inside community between members (HI) Existence of conflicts inside community with organized groups (HI) Existence of conflicts with state agencies (HI)
Economic	Potential losses in economic activities	Forest productive areas (HI) Agriculture productive areas (HI) Industrial and service activities(HI)
	Disruption of livelihoods	Loss of livelihoods (permanent, long or short duration) (HI) Families income dependence from agriculture/forest (HI)
Ecological	Ecosystems conservation	Conservation of rare or representative habitats (HI)[1] Permanent loss of "natural" environment (HI) Loss of biodiversity(HI)
	Natural resources	Quality of water resources (HI) Availability of water resources (HI) Soil erosion (HI) Air pollution (HI) Time to recovery (HI)

TABLE 7.3 Susceptibility/Fragility Indicators—cont'd

Dimension	Variables	Indicators
Cultural	Cultural heritage values	Identification of cultural heritage sites (HI)[1]
	Traditional forest practices	Traditional fire use (HD) Cultural practices on forest (HD
Institutional	Legislation constraints	Planning legislation and practices (HI) Top-down decisions and actions (HI)
	State attitude	Conflicts between agencies (HI) Fire agencies prescriptive (HI) Bureaucracy (HI) Disempowerment (HI)
	Fire service evaluation of property defensibility	Houses defensibility (HI) Forest defensibility (HD)

HD—Hazard dependent indicator; HI—Hazard independent indicator.
[1]*Notes: The underlined indicators are the ones used in the elaboration of a forest fire vulnerability map at the municipal scale in Portuguese test case.*

through indicators (i.e., *forest areas slope, species flammability, forest accesses, and fuel breaks*) which translate the facility for fire outbreak, fire behavior (i.e., intensity and spread velocity), accessibility for fire suppression (e.g., favor the placement of fire brigades which contribute to reduced fire size), and difficulties to fuel management. For instance, a forest on steeper slopes is more vulnerable due to the difficulty of fuel management and suppression tasks. In an area with high flammable vegetation the spread of a fire is easier.

Reducing *buildings susceptibility* in WUI, which are frequently threatened by firebrands and spot fires instead of linear fire fronts (Caballero, 2004), is an opportunity to decrease the occurrence of catastrophic fires. Structural ignitability, rather than wildland fuels, is the principal cause of structural losses during WUI fires (Cohen, 2002).

The roof flammability, building's materials, design and maintenance in relation to its immediate surroundings determine *building's ignition potential* (Cohen, 2012). *Housing security* is more related with preparedness measures to protect the house (e.g., defensible space, fire suppression devices, avoid high flammable vegetation on hedges; flammable materials stored outside the houses) (Cohen, 2012; Collins, 2005).

7.4.2.2.2 Social

The *population characteristics* which better explain susceptibility/fragility are *age* (chronological or functional age), *people disability*, and *one-person households*. These characteristics affect the capability of fuel management, response to a fire outbreak, and personal self-protection or evacuation.

Deprivation and social exclusion is not only related with poverty as a lack of income but associated with multiple types of deprivations (e.g., education, employment, health, housing, access to services, and access to information) with impacts on fire prevention, preparedness, response, and recovery.

People vulnerability is also associated with the level of *integration in communities* which can be translated by indicators like: (1) the *temporary residents and holidaymakers* who, frequently, avoid all the contact with the local community which results in a lack of locally relevant knowledge and interaction with local people vital to mitigate forest fire risk (Gilbert, 2005; Akama and Ivanka, 2010); (2) the *in-migrants in country for less than 2 years* because they could be unfamiliar with the risks of fire in the area; (3) the *ethnic minorities' integration which* contributes to social vulnerability namely through the lack of access to resources and social marginalization. It affects either the cultural management of fire (Wallace, 1970) or preparedness and recovery.

The variable *absence or lack of motivation/opportunity* tries to capture how land and house owners and buildings' residents absence affects fuels management and preparedness measures to reduce houses ignition potential.

The *community fragmentation* translates existing conflicts that could lead to the intentional fire ignition. The roots of these conflicts are related namely with resources exploitation (e.g., hunting, pasture renovation, agricultural cleaning practices, common lands management, nature conservation regulation).

The identification of *people with mental disorders* may be useful because they might be connected with fire risk behaviors or intentions to deliberately light a fire. Severe fires also affect *cardiovascular and respiratory diseases* (Johnston and Bowman, 2005). At the same time, the anxiety affects the capability to cope and to recover from a fire.

7.4.2.2.3 Economic

The economic vulnerability is related with the *potential losses in economic activities* considered through losses of production and/or jobs. It is fundamental to identify the location of large investments in order to develop prevention measures and a quick response in the case of fire ignition.

Sometimes, the losses are small at the community level but they represent the *disruption of livelihoods* that affect deeply the families' income from agriculture and forest.

7.4.2.2.4 Ecological

Two main variables were identified: (1) *ecosystems conservation* characterized by the presence of rare species contributing significantly to the maintenance of biological diversity; (2) *natural resources* availability and quality and *time to recovery*.

7.4.2.2.5 Cultural

Cultural susceptibility/fragility is related with the identification of *cultural heritage sites* (to support their protection and mitigate potential damages), the identification of areas that tourists can use for cultural and recreation practices (e.g., nature observation, picnicking), and the knowledge of several *traditional fire use* (i.e., understanding how, when, and why local communities use and manage fire) (Ganz and Moore, 2002).

7.4.2.2.6 Institutional

Institutional vulnerability is related with *legislation constraints, state attitude*, and the lack of the promotion of a *fire service evaluation of property defensibility*. The current top-down approach defines several procedures, which are not adapted to the social and geographical characteristics of many places, in order to produce an effective, sustained, and integrated risk and disaster management.

The lack of cooperation between fire agencies and communities resulting from a self-centered behavior limits the possibility to find better measures to face the structural roots of vulnerability and to implement the findings from scientific research and lessons learned from past fire events.

7.4.2.3 Lack of Resilience Indicators

It is also likely that exposure and susceptibility/fragility would fall with higher resilience (i.e., capacity to anticipate, to cope, and to recover) (Table 7.4).

7.4.2.3.1 Capacity to Anticipate

Capacity to anticipate is developed through two vectors. Firstly, as there is a big difference between fighting a large fire or one on its early stages, an early detection of a fire requires good *fire surveillance* coverage. To decrease the probability of damaging fires, *fuel management* (e.g., *use of prescribed burning, controlled grazing*) is the most efficient procedure. Concerning the high costs of fuel management it is fundamental to elaborate a *strategic plan* which defines areas where fuel management is more necessary and effective.

Secondly, enhance households' (and also communities') competencies and resources to respond to a fire recognizing the incapacity of fire agencies to protect all people and houses. In order to do it, in the safest conditions, it is essential for the development of *Public awareness campaigns* directed to target groups to enhance *community response preparedness* and *households' preparedness*. Even though the existence of information is necessary it is not the sufficient condition to motivate actions that reduce fire risk (Paton, 2006); therefore, *community engagement* is required. This process looks for the accommodation of diversity and the multiple interests of the community members by facilitating their capacity to forge and improve connections between individuals, organizations, and networks (Chaskin, 2001; Paton, 2011). Several facets of communities (e.g., *sense of community, place attachment, collective efficacy*) facilitate the

TABLE 7.4 Lack of Resilience Indicators

Dimension	Variable	Indicators
Capacity to anticipate	Fire surveillance	Visibility from surveillance lookout towers (HD)[1] Mobile surveillance coverage (HD)
	Fuel management	Defensible space around settlements (HD) Defensible space around buildings (HD) Use of prescribed burning (HD) Strategic plan to manage fuels (HD) Controlled grazing (HD)
	Community engagement	Collective efficacy (HI) Social networks (HI) Local volunteers (HI) Participating planning (HI) Sense of community (HI) Place attachment (HI)
	Public awareness campaigns	Awareness campaigns to target groups (HI)
	Community response preparedness	Community emergency plan (HI) Empowering physical settings (HI)
	Households preparedness	Knowledge about fire safety measures (HD) Adoption of fire safety measures (HD) Competencies and resources to respond to fire (HD)
Capacity to cope	Availability of physical resources	Distance of water points (HD)[1] Distance of fire-fighter brigades to the settlements (HI)[1]
	Community isolation	Road network access (HI)[1]
	Fire ecosystem dependency	Fire dependent ecosystems (HD) Fire sensitive ecosystems (HD) Fire independent ecosystems (HD) Fire influenced ecosystems (HD)
	Fire suppression resources and capabilities	Adequacy of technical and human resources (HI) Training of fire-fighters teams (HI)
Capacity to recover	Natural resources	Vegetation response capacity (HD)
	Insurance	Household insurance (HI) Forest insurance (HI)
	Deprivation and social exclusion	Deprivation index (HI)
	State or other institutions support	Preestablish state or other institutions support funds (HI) Occasional and uncertain state or other institutions support funds(HI)

HD—Hazard dependent indicator; HI—Hazard independent indicator.
[1]*Notes: The underlined indicators are the ones used in the elaboration of a forest fire vulnerability map at the municipal scale in Portuguese test case.*

obtainment of information and material support (Kuhlicke et al., 2011), self-organization, and motivation for collective action (Cheng et al. 2003; Sandström, 2009; Brunckhorst, 2010) which influence their engagement in forest fire prevention and preparedness.

7.4.2.3.2 Capacity to Cope

To control fire, the efficacy in the initial and extended attack is fundamental to decrease vulnerability. The capacity to cope is determined by the *availability of physical resources, fire suppression resources, and capabilities* (Table 7.4).

Community isolation, measured by the *road network access*, favors vulnerability by difficulty the possibility of evacuation in the outbreak of a fire as well as the external support from fire-fighters brigades.

Finally, *fire ecosystem dependency* reveals four categories of ecosystems response to fire (Myers, 2006) and explain the damages in the ecosystems.

7.4.2.3.3 Capacity to Recover

The capacity to recover is related to the households' financial resources (i.e., *insurance* and *deprivation and social exclusion*) as well as the *state or other institutions support* which can be previously established or be denied only in the case of very severe events.

The *vegetation response capacity* translates the capacity to recover the same composition and structure after being affected by a fire. Natural regeneration after fire is very dependent on the fire regime (intensity, frequency, and season) (Lloret et al., 2003).

7.4.2.4 Indicators Implemented in Portuguese Test Case

From the presented list, we selected the indicators more adequate to the geographical characteristics of the test case in collaboration with operational level stakeholders. Accurate data were found only for 14 indicators which clearly limited the comprehensive assessment of forest fire vulnerability. These indicators were mapped using GIS tools and validated in Portuguese test case. The next step was the integration of indicators by dimension. Then we proceed with the aggregation of dimensions and finally the integration of the three components in order to obtain a final score.

The main challenge was deciding on the weight of each indicator, dimension, and component. For the specific test case using the stakeholders experience we proposed that:

> **Exposure index** = 1 × (Population Density) + 3.25 × (Distance of each settlement to the forest) + 3.75 × (WUI characteristics) + 2 × (Distance to extensive grazing areas) (Figure 7.11).

Concerning the susceptibility/fragility and lack of resilience indices to overcome the lack of information at an adequate scale, we need to proceed with research and get our own primary data.

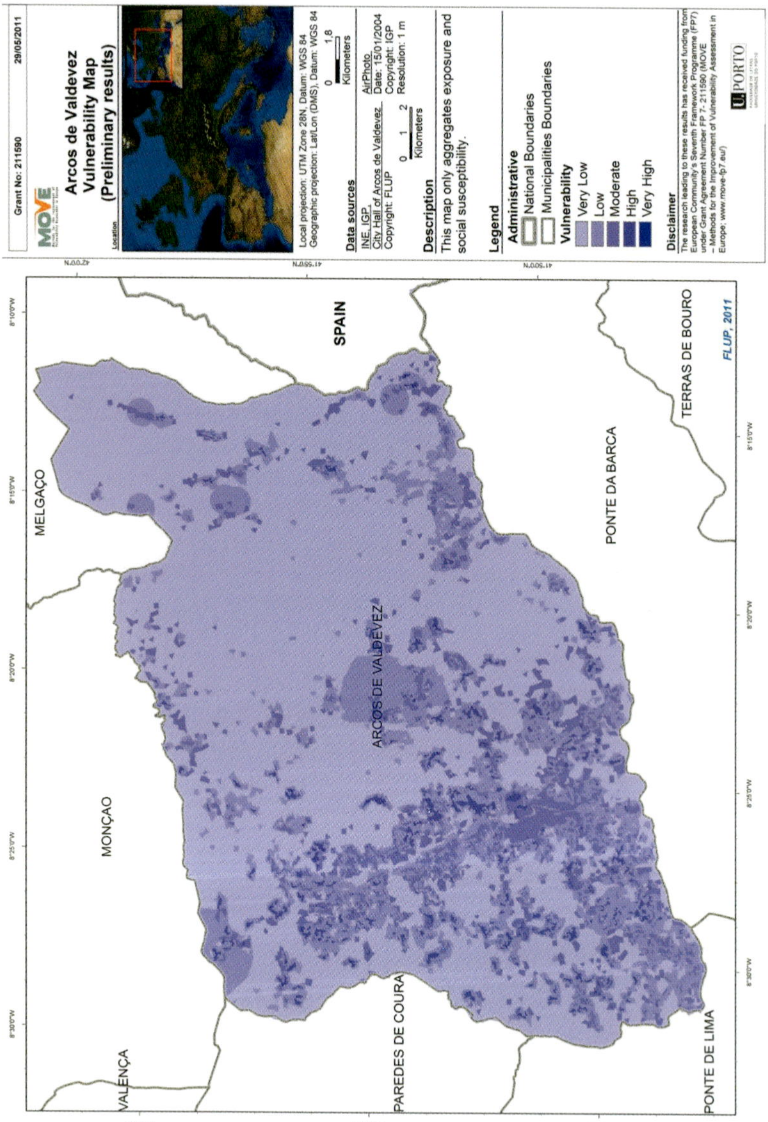

FIGURE 7.11 Vulnerability map of Arcos de Valdevez municipality.

The final vulnerability index should be an aggregation of exposure, susceptibility/fragility, and lack of resilience. It should be highlighted that the fundamental question continues to center on weighting factor to be assigned to each component and this calls for more discourse and research.

7.5 RESULTS

In Portuguese case study, we dealt with two hazards with very different characteristics. Coastal erosion is an oscillatory process around an average position of the coastline, with erosional events. It is, in itself, a reversible process at some time scales. Contrarily, most forest fires have an anthropogenic origin being one of the most predictable hazards.

In the Portuguese case study, three components of vulnerability defined by MOVE's framework were tested. If this framework identifies components and dimensions, on the validation process we identified variables and indicators able to give a comprehensive perspective of the roots of vulnerability.

For both hazards, different approaches were used to determine not only the hazard type (e.g., return period in the case of forest fires and time expectancy in the case of coastal erosion) but also the desirability to show how MOVE's model operationalization can be done using different methods (e.g., in coastal erosion resilience indicators for each dimension were classified into five different types, see Table 7.1, which was not adopted in the case of forest fires).

For both cases, it is possible to use a single index with a numerical value attached to each individual asset, as well as to assign vulnerability assessment at different spatial scales or at different administrative units.

In coastal erosion, the vulnerability results from the sum of the alteration of the agriculture, urban, and ecological functions of the community and it is measured using the percentage of the lands affected by erosion relative to the total surface area affected by each function.

In forest fires, each indicator selected has meaning by itself and can support vulnerability reduction actions, even though all indicators can be aggregated by dimension and component until a single value is obtained to express the overall vulnerability. Some of the indicators selected for coastal erosion and forest fire vulnerability assessment are hazard independent and can be shared by several hazards. The indicators lists presented for both test cases try to cover all the vulnerability roots, however some variables can be measured using another indicators according to the scale of analysis, data availability, and environment and social characteristics of the area. This reflects some flexibility on the framework operationalization without negative impacts in vulnerability assessment.

The stakeholders considered that MOVE's framework is comprehensible but a little complex probably because they were not familiar with the vulnerability concept (Figure 7.12). They expressed their interest in the model implementation but recognize that the use of it requires the approval of the ICNF and ANPC.

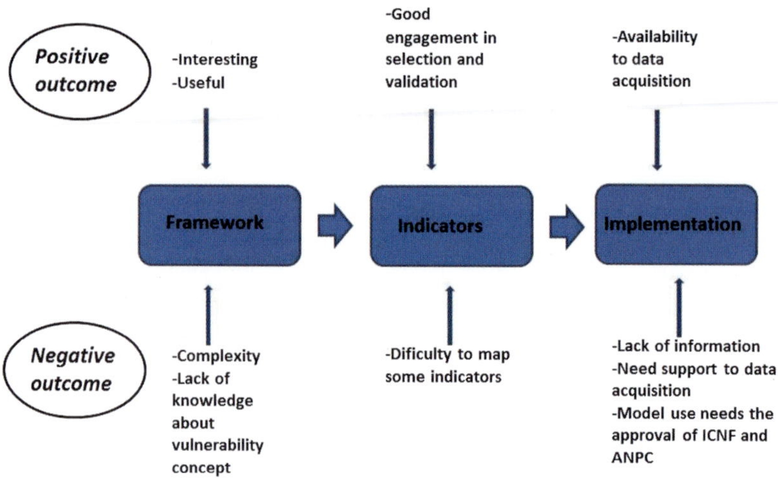

FIGURE 7.12 Stakeholder's opinion on the MOVE's framework and implementation process in Portuguese test case.

7.6 DISCUSSION

In Portugal, the concept of vulnerability was unfamiliar among most of the stakeholders with key responsibilities in risk management. However, a comprehensive assessment of vulnerability is fundamental to promote disaster risk reduction. For example, despite the increase in forest fire suppression efforts, the burned area remains significant. Even though these investments have limited to a certain extent the damages, they did not solve the problem in a sustainable way. The vulnerability assessment seems an appropriate tool to understand when fire is improving or reducing ecosystems condition as well as when fire is threatening social and economic systems or, contrarily, represents benefits.

One of the more interesting aspects of the approaches used is to identify the variability both quantitatively and qualitatively of vulnerability. These approaches are not only focused on the development of an index but can be used to capture the characteristics of vulnerability. If an aggregated index of vulnerability seems to be ideal for the implementation of management measures, the failure to identify the vulnerability's dimensions that cannot be presented in measurable units makes it difficult to interpret (Miller and Ager, 2012). The operationalization of MOVE's framework generates on the one hand a composite index as a global measure of vulnerability, on the other hand each indicator, variable, dimension, and component analysis has its own meaning and produces clear evidences of vulnerability measurements.

The other interesting aspect is that both approaches can be applied to each asset or from household to national scale.

MOVE's framework is a good support for vulnerability assessment. The stakeholders involved in all the research process understood very well the generic framework even if they showed some difficulties in understanding its operationalization. They recognized its importance and utility in risk management (e.g., prevention and preparedness practices, disposal of suppression means, and even combat organization). During the last workshop, the main concern of the participants was about the weighting assigned to each indicator. They are familiar with attributing a value to each indicator and apply it in different social and environmental contexts and for a long period of time. In fact, there are some physical indicators (e.g., vegetation type, slope, distance to forest, or from coastline) that are very similar independently of the location. However, the diversity of social characteristics informs us to not identify weightings independent of the assessment process conducted at a specific location. This means that an assessment process should be undertaken at the scale of analysis in order to identify the weightings applied to the indicator in each locality. Also this process should be repeated periodically to update knowledge of weightings to accommodate changes in social and economic network characteristics over time and/or as a result of resilience intervention (Paton and Tedim, 2012).

The pilot study we realized with experts from different countries, as described in Section 7.3.2, show that the importance assigned to each variable/indicators by the questionnaire respondents were very variable. This can be explained by experts' different background and experience but is also related with geographical and cultural context. Even the four Portuguese experts with field experience and similar activities in forest fire management evaluated the importance of each variable/indicators in different ways. This finding raised some reflections on the validity of this methodology because expert's opinion is conditioned by several subjective judgments namely as a result of their experience and beliefs.

7.7 CONCLUSIONS AND RECOMMENDATIONS

The implementation of MOVE's vulnerability conceptual framework is important to enhance coastal erosion and forest fire risk management in Portugal. If the factors that influence vulnerability are understood, state agencies will be in a better position to make informed choices regarding risk reduction and how the communities can be mobilized to collaborate on it.

The development of MOVE's vulnerability framework revealed the importance of a strong framework to support the comprehension of the vulnerability roots, as well as distinctive operationalization adequate to different geographical contexts and scales, instead of only be focused in the production of a vulnerability index. At the same time, it clearly points out that it is necessary to produce more data in order to make optimum and sustainable use of this framework.

The experience and the interactions with stakeholders during the project were very positive because most of them recognized the importance of improving the assessment of vulnerability. The exchange would enable them to factor new concepts and ideas within their routine and procedural tasks in their risk management activities. However, this can only be done when the Portuguese agencies with responsibilities in risk management (ICNF and ANPC) change their view of vulnerability concept from the degree of damages to a more predictive one as proposed by MOVE's framework. This acceptance will open an opportunity to start building a new generation of risk maps able to support a more dynamic risk reduction.

REFERENCES

Akama, Y., Ivanka, T., 2010. What is community? Facilitating awareness of community through playful triggers. In: Bodkier, K., Bratteteig, T., Loi, D., Robertson, T. (Eds.), PDC 2010: Participation, The Challenge. ACM, New York, USA, pp. 11–20.

Bovio, G., Marzano, R., Aguado, I., Chuvieco, E., Martinez, J., Nieto, H., Salas, J., Berolo, W., Carrega, P., Fox, D., Geronimo, N., Robin, J.G., Gomez, I., Martin, P., Martinez-Vega, J., Vilar, L., Gitas, I., Karteris, M., Tsakalidis, S., 2006. Towards a Euro-mediterranean Wildland Fire Danger Rating System. EUFIRELAB D-08–06: Euro - Mediterranean Wildland Fire Laboratory. A "Wall-Less" Laboratory for Wildland Fire Sciences and Technologies in the Euro-mediterranean Region.

Brunckhorst, D., 2010. Using context in novel community-based natural resource management: landscapes of property, policy and place. Environmental Conservation, 1–7.

Caballero, D., 2004. Conclusions of the Third WARM Workshop on Forest Fires in the Wildland-urban Interface in Europe. WARM Project. Final Report. European Commission, Madrid. Spain.

Chaskin, R., 2001. Building community capacity: a definitional framework and case studies from a comprehensive community initiative. Urban Affairs Review 36 (3), 291–323.

Cheng, A.S., Kruger, L.E., Daniels, S.E., 2003. 'Place' as an integrating concept in natural resource politics: propositions for a social science research agenda. Society and Natural Resources 16 (2), 87–104.

Cohen, J.D., 2002. Wildland-urban Fire- a Different Approach. USDA Forest Service. Fire Sciences Laboratory. Retrieved from: www.firelab.org.

Cohen, J.D., 2012. Residential fire destruction during wildfires: a home ignition problem. In: Spano, D., Bacciu, V., Salis, M. (Eds.), Modelling Fire Behaviour and Risk. Costantino Sirca, pp. 212–217. Available at: http://desa.uniss.it/Modelling%20Fire%20Behaviour%20and%20Risk.pdf.

Collins, T.W., 2005. Households, forests, and fire hazard vulnerability in the American West: a case study of a California community. Environmental Hazards 6, 23–37.

Ganz, D.J., Moore, P.F., 2002. Living with fire. In: Communities in Flames, Proceedings of 1st International Conference on Community Involvement in Fire Management. FAO RAP Publication 25.

Gilbert, J., 2005. An Evaluation of the Street FireWise Community Education Program in the Blue Mountains. Bushfire CRC, New South Wales. pp. 1–3.

Gill, A.M., Moore, P.H.R., 1998. Big versus small fires: the bushfires of greater Sidney, January, 1994. In: Moreno, J.M. (Ed.), Large Fires. Backhuys Publishers, Leiden, Netherlands.

Hinkel, J., 2011. Indicators of vulnerability and adaptive capacity: towards a clarification of the science–policy interface. Global Environmental Change 21, 198–208.

ICNF, 2012. Estatística - dados sobre incêndios florestais. ICNF, Lisboa. Available at: http://www.icnf.pt/portal/florestas/dfci/inc/estatisticas.

IGEO, 2007. Carta de uso e ocupação do solo de Portugal Continental para 2007. Available at: http://www.igeo.pt/e-IGEO/egeo_downloads.htm.

INE, 2012. Os Censos 2011 XV Recenseamento Geral da População e V Recenseamento Geral da Habitação. INE, Lisboa.

Johnston, F.H., Bowman, D.M.J.S., 2005. Fire, landscape and health. In: Fitzpatrick, L., Gregory, G. (Eds.), Proceedings of the 8th National Rural Health Conference. National Rural Health Alliance, Alice Springs, Northern Territory, pp. 1–11.

Keane, R.E., Agee, J.A., Fulé, P., Keeley, J.E., Key, C., Kitchen, S.G., Miller, R., Schulte, L.A., 2008. Ecological effects of large fires on US landscapes: benefit or catastrophe? International Journal of Wildland Fire 17, 696–712.

Kuhlicke, C., Scolobig, A., Tapsell, S., Steinführer, A., De Marchi, B., 2011. Contextualizing social vulnerability: findings from case studies across Europe. Natural Hazards 58, 789–810.

Lloret, F., Pausas, J.G., Vilà, M., 2003. Responses of Mediterranean plant species to different fire frequencies in Garraf Natural Park (Catalonia, Spain): field observations and modeling predictions. Plant Ecology 167, 223–235.

Loureiro E.J.S.M., 2006. Indicadores geomorfológicos e sedimentológicos na avaliação da tendência evolutiva da zona costeira. Aplicação ao concelho de Esposende (Tese de Doutoramento). Universidade do Minho, Braga, p. 322.

Lutz, J.A., Key, C.H., Kolden, C.A., Kane, J.T., van Wagtendonk, J.W., 2011. Fire frequency, area burned, and severity: what is a normal fire year? Fire Ecology 7 (2), 51–65. http://dx.doi.org/10.4996/fireecology.0702051.

Miller, C., Ager, A.A., 2012. A review of recent advances in risk analysis for wildfire management. International Journal of Wildland Fire. 22 (1), 1–14. http://dx.doi.org/10.1071/WF11114.

Myers, R.L., 2006. Living with Fire: Sustaining Ecosystems and Livelihoods through Integrated Fire Management. The Nature Conservancy, Arlington, VA, USA.

Paton, D., 2006. Disaster resilience: integrating individual, community, institutional and environmental perspectives. In: Paton, D., Johnston, D. (Eds.), Disaster Resilience: An Integrated Approach. Charles C. Thomas, Springfield, Ill.

Paton, D., 2011. Community engagement, empowerment and community development. In: The Workshop- Enhancing Sustained Resilience to Natural Risks. The Imperative of Changing the Paradigm. Ponte da Barca. Portugal, June 9, 2011.

Paton, D., Tedim, F., 2012. A dimensão social dos incêndios florestais: identificação de factores que contribuem para uma gestão sustentada e integrada. In: Tedim, F., Paton, D. (Eds.), A dimensão social dos incêndios florestais: para uma gestão integrada e sustentável, Estratégias Criativas, pp. 14–30.

Sandström, C., 2009. Institutional dimensions of comanagement: participation, power and process. Society and Natural Resources 22, 230–244.

Stewart, S.I., Radeloff, V.C., Hammer, R.B., Hawbaker, T.J., 2007. Defining wildland-urban interface. Journal of Forestry 105, 201–207.

Tedim, F., Remelgado, R., Borges, C., Carvalho, S., Martins, J., 2013. Exploring the occurrence of mega-fires in Portugal. Forest Ecology and Management 294, 86–96.

Wallace, B.J., 1970. Hill and Valley Farmers – Socio-economic Change Among a Philippine People. Schenkman Pub. Co, Cambridge, Mass.

Chapter 8

Vulnerability to Heat Waves, Floods, and Landslides in Mountainous Terrain: Test Cases in South Tyrol

Maria Papathoma-Köhle*, Thorsten Ulbrich**, Margareth Keiler†, Lydia Pedoth¶, Reinhold Totschnig§, Thomas Glade*, Stefan Schneiderbauer¶, Unni Eidswig‡

*Department of Geography and Regional Research, University of Vienna, Vienna, Austria, †Institute of Geography, University of Bern, Bern, Switzerland, ‡NGI, Oslo, Norway, §eb&p Umweltbüro GmbH, Klagenfurt, Austria, ¶Institute for Applied Remote Sensing, European Academy of Bolzano (EURAC), Bozen/Bolzano, Italy, **Institute of Meteorology, Department of Earth Sciences, Freie Universität Berlin

Chapter Outline

Assessment of Vulnerability to Natural Hazards. http://dx.doi.org/10.1016/B978-0-12-410528-7.00008-4

8.1 INTRODUCTION

South Tyrol is located in the Eastern Alps, south of the main chain. The climate is mild and dry and the average temperature is 12.3 °C. It includes mainly the upper part of the Etsch/Adige river catchment and its contributors. The landscape is characterized by high mountains (up to 3905 m a.s.l.) and densely populated and intensively used valleys. South Tyrol has a population of 500,000, which is split into three different language groups (German, Italian, and Ladin). South Tyrol is part of the autonomous region Trentino-Alto Adige and, therefore, has a wide independence in legislation. The present study focuses on the assessment of different dimensions of vulnerability to landslides, floods, and heat waves at local scale in South Tyrol (Table 8.1). In this chapter, the term "landslides" is being used to express all these processes that can be defined as the downslope movement of soil, rock, or debris due to gravitational forces that can be triggered by rainfall, rapid snow melting, slope undercutting, etc. (Crozier, 1999; Glade and Crozier, 2005). The studies were conducted in close contact with the stakeholders that were involved at the early stages in the development of the methodologies (Table 8.1).

All studies were conducted on a local scale in different locations in South Tyrol that can be seen in the map (Figure 8.1).

TABLE 8.1 Hazards That are Investigated in These Case Studies, the Scale of the Study, and the Stakeholders Involved

Hazard Type	Location	Vulnerability Dimension	Stakeholders Involved
Landslides	Martello (Martell)	Physical	Local authorities (Hydraulic engineering, Fire and Civil Protection Departments of Autonomous Province of Bolzano)
Floods	Vipiteno (Sterzing)	Social	Residents and local Business, local and provincial administration
Heat waves	Bolzano (Bozen)	Social	Provincial and local health care services, Civil Protection Department, regional and local authorities, residents (especially elderly)

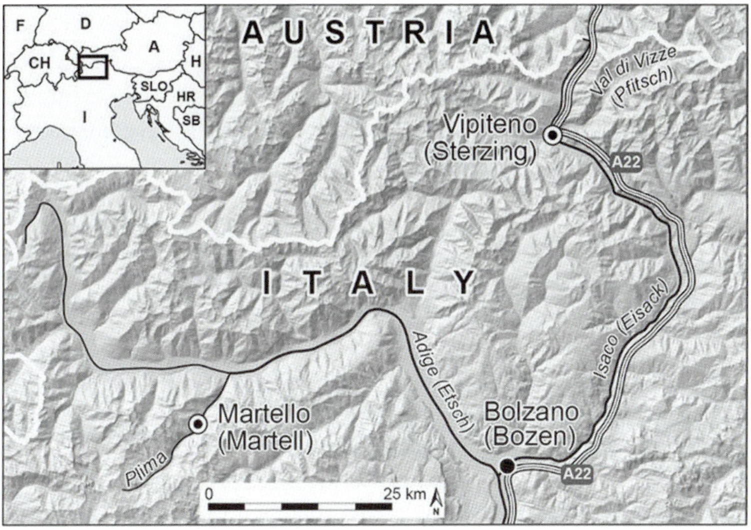

FIGURE 8.1 The location of the three case studies in South Tyrol.

8.1.1 The Effect of Climate Change on Natural Hazards

The report of IPCC (2012) suggests that "There is high confidence that changes in heat waves, glacial retreat, and/or permafrost degradation will affect high-mountain phenomena such as slope instabilities, mass movements, and glacial lake outburst floods. There is also high confidence that changes in heavy precipitation will affect landslides in some regions" (IPCC, 2012, p. 114). More specifically, in South Tyrol, Staffler et al. (2008) identified alpine torrents and river catchments in which future climate change may modify the hazard situation as far as floods and debris flow are concerned. The study of Staffler et al. (2008) revealed that an increase in the intensity and frequency of flood and debris flow events in the area is to be expected and they pointed out the importance of reducing the vulnerability of the elements at risk in order to reduce overall risk.

As far as heat waves are concerned, time–series measurements of land-surface and sea surface temperature of the last century show a clear warming trend. A shift in mean air temperature may lead to a nonlinear response in the frequency of extreme events such as heat waves (IPCC, 2001). Actually, recent regional climate models suggest an increased variability of temperature for the future in addition to an increase in mean temperature, which would likely augment the frequency of extreme events (Schär et al., 2004). In South Tyrol, the average temperature in the summer of 2003 was about 3–4° above the average of the last four decades (APB/SHA, 2003). However, predictions for the future also suggest an increase in the social susceptibility at local level, due to an increase in the number and percentage of elderly people.

8.1.2 Local Preparedness in South Tyrol

The province of South Tyrol has produced a publicly available Internet platform (Hazard Browser Südtirol[1]), showing information on catastrophic events related to mass movements and floods. At municipality level, local hazard maps are also available together with a report describing the methods, models, definitions, and software that has been used for hazard mapping. An early warning system for landslides in two case study areas in South Tyrol has been developed within the project ILEWS (Integrative Landslide Early Warning Systems) (Röhrs and Glade, 2010). In the case study area for landslides (Martell) an early warning system that is connected to the neighboring reservoir has been established and response exercises are regularly carried out. Following the catastrophic event of 1987 there was permanent removal and relocation of buildings, as well as prohibition of building in areas previously built at the bank of the river. Moreover, a protective reinforced concrete wall was built at the bank of the river in parts of the valley (Pfitscher, 1996).

Regarding floods, structural measures for flood protection are installed on the Eisack River and on most of the contributors upstream of the case study Sterzing/Pfitsch. Nonetheless, a flood hazard map indicates a high level of exposure to flood risk for a big part of the site (see Figure 8.2). A major river dredging was implemented to reduce the risk level. The provincial warning system is based on sirens combined with news in local broadcast stations. Early warning systems are based on weather information and a precipitation discharge model. Emergency exercises are held regularly in the autonomous province of Bolzano. In November 2005, in Sterzing and Pfitsch a big intercommunity emergency and simulated evacuation on a flooding/debris flow scenario was exercised.

Regarding heat waves, following the summer of 2003 the national civil protection department set up a national heat health warning system (HHWWS) for some Italian cities including Bolzano, which monitors the heat health danger situation. In addition, the local authorities of the autonomous province of Bolzano have set up a regional forecast system and local initiatives for elderly people in case of heat waves. The lack of consultation between the authorities at local and national level often leads to different results. Consequently, the warnings do not always coincide. The activities carried out by the local administration focus on short-term measures that help to reduce the direct impact of emerging heat. Long-term measures to mitigate impacts such as the avoidance of urban heat islands or the creation of fresh air corridors are currently foreseen (or are under discussion) within the context of urban planning.

Given the expected changes regarding the intensity and frequency of some natural processes due to climate change in combination with gaps in local preparedness and the need for reliable information to support risk reduction strategies, the assessment of different dimensions of vulnerability in the area is considered essential.

1. Available at: http://www.provinz.bz.it/wasserschutzbauten/wildbachverbauung/hazardbrowser.asp.

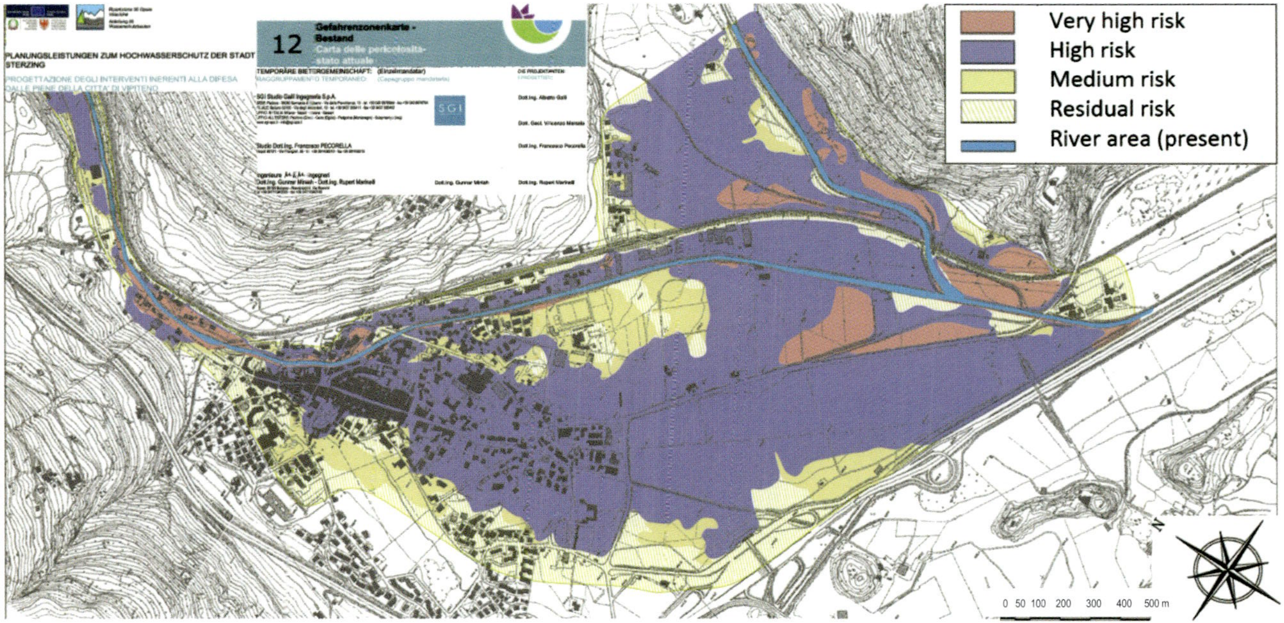

FIGURE 8.2 Great potential flood damage derives from situating big parts of the settlement in zones with high risk. In particular, the historic city center is highly exposed, which is essential for economy and tourism. Also other critical infrastructure such as transportation lines are situated in high risk zones. *Autonomous Province of Bolzano.*

In this chapter, social and physical vulnerability to three types of hazards is assessed and the results are demonstrated through the three following case studies.

8.2 LANDSLIDES

8.2.1 The Study Area: Martell

The case study area (Martell) has suffered in the past from flood and debris flow events. However, the most disastrous event that has been recorded in the area until now, and it is also the one that we are using in our case study, is the event of August 1987, which was a combination of natural and man-made disaster. The height of water and debris reached at some areas in the valley the height of 2.5–3 m and the destruction was severe. At least 12 houses were totally swept away and many more were severely damaged (Pfitscher, 1996). Roads were washed away, animals died, industry and agricultural buildings suffered damages and loss of equipment, water, sewage, and electricity lines were destroyed. Fortunately, due to early evacuation there were no casualties (Pfitscher, 1996). The cost of the specific event for the Municipality of Martell totaled 7,762,000,000 Italian Lire (Pfitscher, 1996), which corresponds to approximately 8.5 million € in indexed 2011 values.

8.2.2 Vulnerability Assessment Methodology for Landslides

This case study assesses physical and economic vulnerability to debris flow. However, the results of the case study can be used in the future for risk management decisions including risk reduction (exposure and vulnerability reduction), prevention, and mitigation. Physical vulnerability is often defined as the degree of loss. In more detail, natural scientists define physical vulnerability mainly as: "the degree of loss to a given element or set of elements within the area affected by a hazard. It is expressed on a scale of 0 (no loss) to 1 (total loss)" (UNDRO, 1984). Although the physical vulnerability of an element at risk depends on a series of indicators and it is more complicated than a damage assessment, the degree of loss following a disastrous event bears information regarding the vulnerability of an element at risk. The indicator used in our methodology is the degree of loss. It depends on two values: the monetary loss and the object value. Our focus was to collect information regarding the monetary loss and the value of the elements at risk (the buildings in our study area).

The aim of our study is to design a vulnerability curve as a function of the intensity of the process and the degree of loss. The degree of loss equals the monetary damage of an element expressed as the percentage of its overall value. The curve was based on data concerning the event of 1987 in the Martell valley. A stepwise description of our methodology follows:

1. *Data collection*: photos of the buildings, damage information, and information on previous events and existing compensation data for specific buildings.
2. *Intensity assessment*: from photos of damaged buildings following an event the height of the debris deposits can be assessed.

3. *Monetary damage assessment*: the damages could be evaluated from photos and with the help of price lists indicating the costs of repair/reconstruction the monetary damage can be calculated. Information regarding the costs of repair of buildings following a debris flow was taken from Kaswalder (2009) and it was used throughout the study with some necessary modifications. The monetary loss includes only the costs of direct losses (cleaning, new paint on the walls, new doors and windows, testing and reinstalling electricity, heating and sewage systems, etc.).
4. *Calculation of the degree of loss*: The degree of loss of each building was calculated as the monetary damage expressed as percentage of the overall value of the building. The object value was calculated using the real compensation price for rebuilding (Italian Lire of 1989 was converted to Euros 2011).
5. *Development of the curve*: Based on the intensity and degree of loss for each building the vulnerability curve could be developed.
6. *Validation of results*: the results can be compared to real compensation data.

The resulting curve can be improved in the future by adding more data of more buildings and it can be used for loss estimation of future events. In order to ensure high quality of damage information, the post-event damage assessment methods have to be improved. The vulnerability curve can be integrated in a tool The tool allows rapid damage assessment and loss estimation while at the same time the new information that will be included will improve the existing vulnerability curve.

8.2.3 Results, Validation, and Discussion on the Landslide Case Study

Following the methodological steps mentioned above, a vulnerability curve was developed (Figure 8.3). The degree of loss was calculated for 51 buildings that were damaged in the event of 1987. The focus of the study was strictly on residential 1, 2, or 3 storey buildings with similar architecture, material, and condition. Photographic documentation provided information not only on the extent of the damage on buildings following specific events but also information on buildings characteristics such as use, material, condition, floors, surroundings, etc. Additional background information included information of historic events and compensation data, existing protection measures, and a GIS database showing the land use, the location of the protection, and the distribution and size of the buildings. The vulnerability curve shows clearly that the higher the intensity of the process, the higher the degree of loss. In order to validate our results, a second curve was plotted using data from the municipality of Martell regarding the compensation that the house owners received following the event of 1987. In this way, the real degree of loss of a number of buildings could be calculated and compared with our own results.

The final product of our study is the vulnerability and the validation curve (Figure 8.3) based on the assessment of the intensity of the process and the monetary damage for individual buildings that were affected.

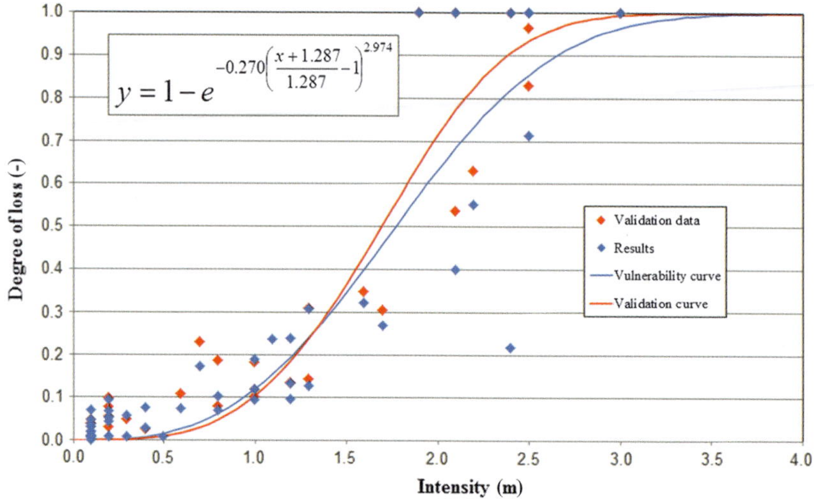

FIGURE 8.3 The vulnerability and the validation curve for buildings affected by the August 1987 event in Martell valley, South Tyrol. *Data source: Municipality of Martell, South Tyrol (Papathoma-Köhle et al., 2012).*

In this study, post-damage photographic documentation was used for the first time in order to calculate the monetary damage and the degree of loss of individual buildings. The vulnerability curve produced showed that total destruction of buildings can occur with intensities less than 2 m. For intensities higher than 1.4 m the degree of loss raises rapidly. This can be explained by the fact that after 1–1.5 m the deposit height reaches the window and it is possible that it enters the interior, which means more damages and costs. However, damages may appear already with intensities of a few centimeters due to damages in the basement. The validation curve that was developed based on compensation data appears very similar to the vulnerability curve based on photographic documentation.

The vulnerability curve that was produced gives the opportunity to decision makers to calculate the costs of future events not only for individual buildings but for the entire municipality. However, in order to improve the existing vulnerability curve or produce more curves for different types of landslides, improvements in the way damage assessment is carried out are necessary. The methodology can be transferred to areas with similar architecture and building material as it is and with modifications it can be transferred in any other places in the world facing the impact of similar hazards. The methodology itself can be slightly modified to be used for other types of landslides (e.g., shallow landslides, rock falls, etc.) but also for snow avalanches. However, the assessment of the intensity of the process on individual buildings may be exceptionally difficult, partly due to lack of data and partly due to difficulties in adequately expressing intensity.

8.2.4 Uncertainty Analysis for the Landslide Case Study

A very important part of the vulnerability assessment is the analysis of the uncertainties involved. Each step of the above described methodology for the development of the vulnerability curve bears a certain amount of uncertainty. In more detail, the sources of uncertainty for the empirical data points are listed below:

- Uncertainties in intensity:
  - The use of deposit height as the single intensity parameter. The effect of velocity and of duration of the water and debris staying inside the building was not considered.
  - Assessment of intensity from debris- and water-marks on the buildings and extent of the damage, interpreted from photographs. The quality, the aspect, and the number of photos per building influence the credibility and preciseness of the intensity assessment.
- Uncertainties in the degree of loss:
  - Assessment of damage pattern from photos which show mainly the external damage and not the interior damage.
  - Assessment of costs of reconstruction which was based on and modified from a list of repair costs that was developed for floods rather than debris flows.
  - Estimation of value of buildings with unknown use and size of basement and attic.
- Credibility of the existing data: For some buildings there is a mismatch between the damage shown on the photo and the received compensation.

To quantify the uncertainty in the proposed model, uncertainty in the empirical data points was assessed. The intensity and the degree of loss were specified as ranges rather than as single values, by using expert judgment and estimations. The ranges were defined by specifying six data points for each observed building.

The empirical data points, with uncertainty, were used to quantify the uncertainty in the vulnerability model for buildings affected by debris flow. The model uncertainty is represented by uncertainty bands (Figure 8.4). The uncertainty bands are presented in terms of percentiles: the 64 percentile means that 64% of the empirical data is below the curve, the 13 percentile means that 13% of the empirical data is below the curve.

Figure 8.4 illustrates the uncertainty in the model and is comprehensive for the stakeholders. The vulnerability curve represents the averaged degree of loss of the observed buildings, while the uncertainty bands represent the spread in the data represented by observations of single buildings. However, the exact probability distribution of damage is not shown and the information could not be applied directly into a detailed probabilistic loss assessment. In order to provide a probability distribution of damage for intensity, the empirical data could also be used to define a fragility function for buildings hit by debris flow.

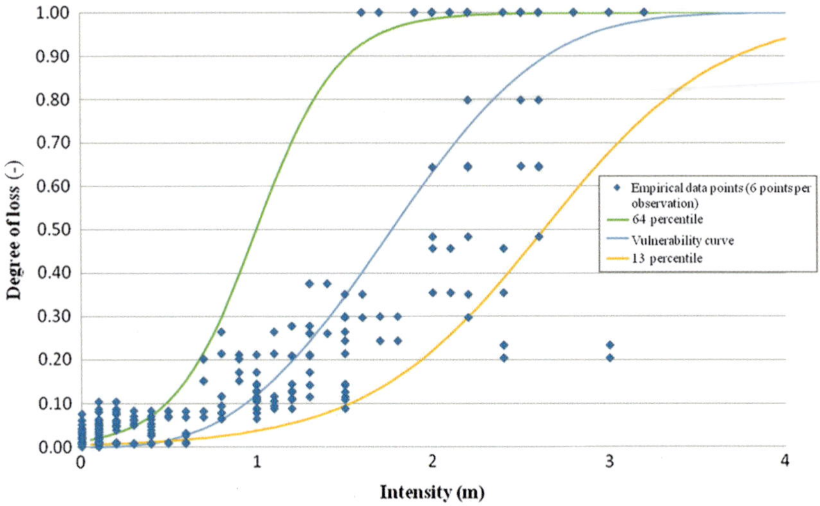

FIGURE 8.4 The model uncertainty represented by uncertainty bands. A high (64% of the empirical data below the curve) and a low (13% of the empirical data below the curve) estimate of the degree of loss were developed.

8.3 FLOODS

8.3.1 The Study Area: Sterzing/Pfitsch

Sterzing and the adjacent settlement of Pfitsch are among the municipalities most prone to be affected by extreme floods. In this area, three mountain torrents—Pfitscher Bach, Mareiter Bach, and Faller Bach—open out into Eisack River within or near built-up areas. Sterzing on the east bank and Pfitsch on the west bank of the Eisack River are located in their flood plain at 950 m a.s.l. The valley is important for transalpine traffic since the Brenner highway (A22) and railway pass through (Figure 8.1). A (preliminary) risk zone map (Figure 8.2) illustrates that a major part of both settlements, including the historic center of Sterzing and critical transport infrastructure is highly exposed.

To protect the settlement, river regulation started in the 1870s, largely reducing the frequency of floods and debris flows. Additional river regulations on the contributors of Eisack in the 1970s further reduced flood frequency. To improve the level of protection the provincial administration dredged the Eisack River in 2011.

8.3.2 Vulnerability Assessment Methodology for Floods

Detecting the susceptibility of social and cultural systems is an essential part of vulnerability assessment. Attitudes and motivations that encourage people to take flood precaution or not may be explored by qualitative methods. In the flood case study, we utilize focus group discussions (FGD) to identify attitudes and beliefs that determine motives and, therefore, influence flood risk behavior and vulnerability.

The first step of applying the methodology is to gain support of the provincial Department for Hydraulic Engineering and the mayors of Sterzing and Pfitsch.

In step two, we collected background information, such as data on historic floods and debris flows, flood protection measures, flood hazard maps (see Figure 8.2), and a report on social vulnerability of Sterzing and Pfitsch (De Marchi et al., 2007).

The third step was to invite local residents and businesses for an FGD through the local monthly magazine "Der Erker" (Der Erker 06.2010). Institutional stakeholders and larger enterprises located in the hazard zone were invited personally.

The fourth step was to build up a set of research questions to identify criteria affecting vulnerability of residents and businesses. We formulated open topics for participants to discuss. A schedule with a brief agenda and discussion topics was provided for participants.

Step five was composing the discussion groups. We assembled the groups with stakeholders from administration and relief organizations mixed with private stakeholders. We planned to conduct three focus group sessions: one with local businesses and two with residents split by different levels of exposure. Owing to low participation we could only hold one mixed session.

In step six the group discussions were conducted. The group included six residents of Sterzing/Pfitsch and one representative of the administration of each Sterzing and Pfitsch, the local fire brigade, and the provincial Hydraulic Engineering Department. Small businesses were also represented.

The discussion was structured into four parts:

- A brief introduction of the context, the research method, and setting up discussion rules.
- A round of introductions focusing on how participants are connected to flooding.
- Open discussion: The questions aimed at triggering discussions on flood awareness, responsibilities for flood protection, private flood precaution, and reaction to a flood warning. The moderators intervened only to steer the flow of the discussion, to dig deeper when participants raised thrilling issues, and to clarify when arguments remained unclear, avoiding inducing bias.
- A short summary by the researchers and a feedback from participants.

With participants consent the session was taped.

Step seven was to hand out a survey composed of 20 questions following the discussion. The survey supports the discussion results by describing the setting in more detail and characterizing the participants' background. We gathered data on: flood awareness, flood experience, private flood control measures, anticipated reaction to flood warning, living conditions, and demographics. Combining qualitative data gathered in the FGD and quantitative data generated in the survey helped reveal, to best effect, which attitudes, beliefs, and motivations lay behind arguments.

Finally, step eight was the analysis of the accumulated data. To conduct the analysis we prepared a transcript of the FGD and searched for arguments related to the following indicators:

- Flood risk awareness: Only if the residents and businesses at risk are aware of the possibility and consequences of a flood are they likely to take action to protect their lives and prevent or mitigate material and other losses.
- Attitudes on responsibilities: Only if residents and businesses understand and know their own responsibility for flood protection measures they might take action.
- Attitudes on self-precaution. Only if residents and businesses know possible measures of self-protection and have a positive attitude toward them, appraising them as reasonable and feasible, they may take action.
- Reaction to warning: When residents and businessmen receive a warning they need to react fast and efficiently to protect their lives and property against flash flooding. Therefore, it is important for them to know in advance how to respond.

We also investigated participants' risk perceptions, flood beliefs, integration in local social networks, and knowledge of the warning system. These data were contrasted with each other, the background information collected on the study site and the survey. The results were published in the local press (Der Erker 08.2010) and a protocol was sent to the participants. The results of the analysis are summarized in the following section. The method can be adapted in the future to investigate also other aspects of risk governance.

8.3.3 Results and Discussion on the Flood Case Study

The FGD was analyzed with regard to vulnerability, focusing on the indicators—"flood risk awareness", "attitude on responsibilities", and "self-precaution" and "reaction to warning", as well as related topics that arose during the discussion. Participants brought up topics on exposure, responsibility of the authorities for flood precaution, emergency response, and evacuation. Another intensively discussed topic was the authorities' plans for flood control measures and the communication thereof, as well as the transparency and scope for participation that was integrated into that planning process.

Participants agreed that "everybody living here next to the river is aware of what will happen, when the river bursts its banks" stating high-risk awareness regarding floods. Contradicting this low participation of the FGD alludes lacking of flood awareness. This lack might be a consequence of having been spared of flooding for a long period.

It seems that existing flood controls support local risk awareness. Nevertheless, several residents doubt the high rating of risk levels. They reason this appraisal from the absence of severe floods after constructing defenses in the 1970s. Based on experiences from past events, residents rely on sufficient warning lead times. In contrast to this, the provincial Hydraulic Engineering Department warns that lead times are maximum 6h and flash floods can surprise at night.

It was remarkable that participants disagreed as well on the scope of self-precaution measures as on attributing responsibility for flood control. Some participants take flood precautions. Those accept private responsibility and say: "residents also have to do something themselves" to protect their own property. This group recognizes that there is no 100% flood security. In contrast, others expect very little effect from private measures and claim: "responsibility clearly lies with the authorities". The fire brigade chief and local mayors advocate this view. They consider it more efficient and fair.

This research revealed twofold effects on vulnerability and lead us to define local factors that are increasing as well as reducing vulnerability. The factors that are increasing or decreasing vulnerability are shown in Table 8.2.

The study involved local stakeholders in risk governance. It improved communication and understanding among stakeholders. Surprisingly, we found that all residents participating in the discussion agree to appreciate public flood control measures, even though they have to give up private land. Nevertheless, they demand to be informed about plans earlier and to be granted more scope for participation.

TABLE 8.2 Factors That Increase or Decrease Social Vulnerability According to the Results of the Focus Group discussions

Factors Increasing Social Vulnerability	Factors Decreasing Social Vulnerability
• High exposure in most areas of the town • Susceptibility of economy and critical infrastructure • Underestimation of risks • Overreliance in existing protection measures • Objectively incorrect assumptions from past experiences • Too much trust in the warning system • Often lack of feeling for own responsibility • Often little knowledge of private precaution measures	• Participants are aware of flood risks, (nevertheless overall participation was low) • Well-understood siren-based warning system • Short routes for evacuation • Some residents take private precaution measures • No oil heating being installed in endangered households (district heating instead) • Functioning local social network • Residents agree on flood defense as a target of spatial planning

Applying FGD enabled us to generate data with relevant stakeholders in a time frame that was easy to schedule and independent from external data providers. Furthermore, the application was fast and cheap and produced relevant findings, which might not have been discovered by quantitative research alone.

Understanding attitudes and beliefs that lead to or prevent taking flood defense actions is a premise for successfully encouraging vulnerable populations to take individual measures for vulnerability reduction and to understand social/cultural vulnerability. FGDs are an adequate and easy way to uncover attitudes and beliefs. Social interaction and group dynamics generate high validity, but they are sensitive to low participation and depend on good group dynamics in discussions. Results are validated by the discussion among stakeholders.

A major drawback of our case study was that a great number of affected local residents and businessmen were not integrated. Based on only 10 participants our findings have limited explanatory power. Yet, in other case studies with more recent floods the method succeeded to attract far more stakeholders.

A motivation of this case study lay in exploring the application of the method for assessing vulnerability. Testing the method has succeeded, since it provided substantial knowledge on aspects raising and reducing vulnerability in our study area. Our findings are backed up by research on social vulnerability conducted by FLOODSITE (De Marchi et al., 2007). However, constructions for an extensive river dredging for flood protection started in 2011. Therefore, the level of exposure has changed.

Additionally, this case study contributed to easing conflicts, finding a compromise between local stakeholders, and in creating mutual understanding and trust. FGD therefore proved to support vulnerability assessment and to be an efficient way to allow public participation and to reduce potential conflicts in local planning.

8.4 HEAT WAVES

8.4.1 The Study Area: The City of Bolzano

Bolzano lies in the center of the south-eastern Alps, at an altitude of about 250 m a.s.l. Due to its location in the basin of a deep valley, the city is affected by high temperatures and heat waves during the summer months. During the last 30 years, the annual average temperature in Bolzano increased by +1.5 °C (see Figure 8.5). However, not only the maximum temperature during the days but also the minimal temperature during the night is relevant to the impacts of heat stress on human health. In Bolzano, the number of tropical nights (T_{min} > 20 °C) has increased significantly over the last 20 years. Until 1995, Bolzano had less than five tropical nights per year, whereas 20 tropical nights were recorded in 2010.

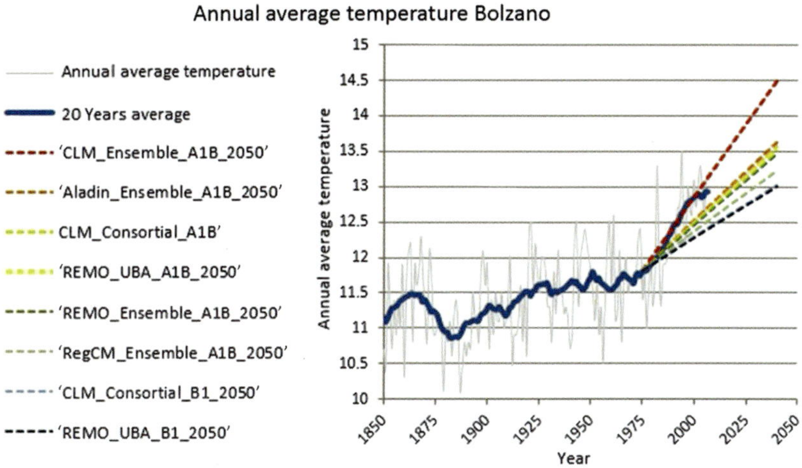

FIGURE 8.5 Warming in Bolzano. The graph shows the increase of the annual average temperature as yearly values (gray line) and smoothed as 20 year average (blue line) in the last 150 years since 1850. *EURAC based on data provided by the autonomous province of Bolzano.*

The colored dashed graphs in Figure 8.5 represent the results of various climate scenarios for South Tyrol. The measured warming from 1975 to present mainly follows the red dashed line of that scenario with the highest values of temperature increase.

8.4.2 Vulnerability Assessment Methodology for Heat Waves

The case study focuses on the social dimension of vulnerability. To establish the impact of heat waves on the citizens of Bolzano, it seemed most reasonable to focus on subgroups that are most susceptible to heat waves. As already identified by previous studies (Drinkwater and Horvath, 1979; Basu and Samet, 2002), these are the elderly.

Spatial exposure and social vulnerability were assessed by using the following methods:

1. Remote sensing (Spatial exposure):

Land-surface temperature was derived from the thermal band of various Landsat Images after atmospheric correction in order to receive rough information about temperature distribution within Bolzano. Overall 12 Landsat 5 scenes of summer months (June, July, August, and September) were used for the period 2003–2009 and in each scene the temperature deviation from each pixel to the maximum temperature value of the same scene was calculated.

2. Statistical analysis (Social susceptibility):

Correlation analysis was carried out by comparing time series of demographic and health-related data with respect to time series of meteorological data. The demographic dataset was split into various subgroups according to age and gender. Hospital emergency/admission data were divided by gender and age in order to focus on different subgroups. Two different age limits were chosen in order to establish a relationship between the climate data and the hospital data for men and women (A) aged 65 and older; and (B) aged 75 and older.

The aim of the statistical analysis was to investigate the correlation between the hospital admissions of the different subgroups and the temperature data in order to find out the most vulnerable population group. A correlation between the two datasets would show the impact of heat stress on the different subgroups and sustain the hypothesis that heat problems are an issue in Bolzano. Therefore, we compared the means of the various subgroups by performing *T*-tests.

3. Potential heat wave index (Social vulnerability):

A composite indicator was applied which combines different factors, using the semiquantitative fuzzy logic technique (Kropp et al., 2009). This approach allowed for an assessment and comparison of the social vulnerability of the five districts of Bolzano, considering spatial exposure and social susceptibility. Quantitative data provided by the province and the regional hospital of Bolzano were analyzed and visualized with the statistical programming tool R and ArcGIS within the scope of identifying the parts of the city which are more vulnerable than other ones, a comparative vulnerability assessment was then made within Bolzano. This technique has the advantage that variable values are allocated gradually rather than in binary fashion. Thus, a ranking of the districts referring to possible heat wave impacts was created.

4. Stakeholder involvement (Lack of resilience, Risk Governance):

Stakeholders such as responsible for heat waves in civil protection and municipality, experts, health care managers, representatives from volunteer organizations, and spatial planning experts were interviewed. The aim of these interviews was to collect information on awareness issues, experience with heat waves in the recent past, existing monitoring and forecast systems, and information flow issues between different institutions but also media and the population. The interviews may enable the identification of gaps and the possibilities for improvement.

The indicators used for each of the above mentioned approaches are shown in Table 8.3.

8.4.3 Results, Validation, and Discussion on the Heat Wave Case Study

Land-surface temperature maps that were produced based on Landsat satellite imagery indicated the city center, as well as the city-adjacent industrial

TABLE 8.3 Methods and Indicators Used for the Assessment of Vulnerability to Heat Waves

Method	Indicators
Statistical analysis	• Maximum air temperature (exposure) • Number of heat days; i.e., days with $T_{max} \geq 30\,°C$ (exposure) • Number of heat waves, i.e., number of events with three consecutive days of $T_{max} \geq 30\,°C$ (exposure) • Number of tropical nights • Number of combined heat days and tropical nights (exposure) • Relative air humidity (exposure) • Combined indicators: the dew point temperature and the heat-index. Both combine temperature and relative air humidity (exposure) (Steadman, 1979) • Elderly population divided by gender, considering different age classes (≥65 years and ≥75 years) (sensitivity indicator)
Potential heat wave index	• Percentage of impervious area (exposure) • Percentage of area with land-surface temperature >28 °C, derived from Landsat images of July 2003 (exposure) • Population density (people/km^2) (sensitivity) • Population older than 65 years (sensitivity) • Population older than 65 years living alone (sensitivity)
Stakeholder involvement	• Risk monitoring and forecasting: analysis of existing monitoring system • Information flow: analysis of the information flow in case of heat waves (responsible authority for the forecast, involvement of mass media, and dissemination through health care institutions) • Response measures for most vulnerable population groups: through interviews with managers of care homes and volunteer associations • Risk governance: assessment of the procedures of risk governance in the case study areas

zone, as the hottest areas of Bolzano. These areas are coinciding with the areas of the highest values of soil sealing. Therefore, the results require a validation check in order to be used in heat wave studies that are mainly dealing with air temperature values. Nevertheless, the land-surface temperature may indicate areas representing an urban heat island effect.

In order to identify the impact of heat waves on the elderly population of Bolzano, time series of climate data and hospital emergency/admission data for the years 2003, 2006, and 2009 were analyzed in parallel and in comparison to each other. By means of simple correlation analysis, the daily numbers of hospital emergencies/admissions were plotted against various exposure factors (maximum daily temperature, relative air humidity, etc.)

and considering a potential time-lag of a heat impact but no relationship could be established.

Boxplots that visualize the distribution of the daily number of hospital emergencies/admissions according to these exposure factors confirmed that there is no extraordinary difference between hotter and less hot days. Nevertheless, with a time-lag of 3 days, for the heat summer of 2003 a slight visible difference in the boxplots could be detected for the hospital data of both women ≥65 years and women ≥75 years between the subgroups of days without heat wave issues and the subgroup which considered only the hospital data for days with a combination of heat day conditions and tropical night conditions (see Figure 8.6(a) and (b)). For the latter, the mean value was visibly greater. However, the result of a Welch's *T*-Test which compares the mean values of the two entities shows that on a significance level of 95% there is no significant difference between the two.

The fact that no Welch's *T*-tests reached the claimed significance level is probably due to the high fluctuation rate in the number of samples per day compared to the low number of samples per day. Therefore, for small cities with such small number of samples it seems to be difficult to show correlations of weather condition and health issues by means of statistical analysis.

Additionally, a potential heat wave impact index was developed for the different districts of Bolzano in order to identify the districts that are most susceptible to heat spells (Figure 8.7). The indicators used are those proposed by Kropp et al. (2009). Problems rose through the large areas of the administrative districts incorporating parts of the cities with a large variety of urban structure and population densities and hence hindering the identification of most endangered areas.

It must be acknowledged that the index values refer to the local reality and, therefore, they have to be interpreted in a comparative way, e.g., as a ranking rather than as absolute values. An index value of 0.98 signifies that the susceptibility in the district Europa/Novacella is relatively higher than in S. Giovanni/Don Bosco or any other district. In the districts Oltrisarco/Aslag and Gries/S. Quirinio the susceptibility is about the same; and the district Centro/Piani di Bolzano is the least susceptible district. This is mostly due to the low fraction of impervious area when looking at the whole district and does not really reflect the situation in the most densely populated part of it.

Heat wave impacts are commonly measured by means of mortality rate which may increase above the expected rate during extreme heat events. Bolzano is a relatively small city with only about 100,000 inhabitants. So, it was much more difficult to clearly identify heat health related issues, since it is not possible to assess the issue by means of mortality rate as in bigger cities.

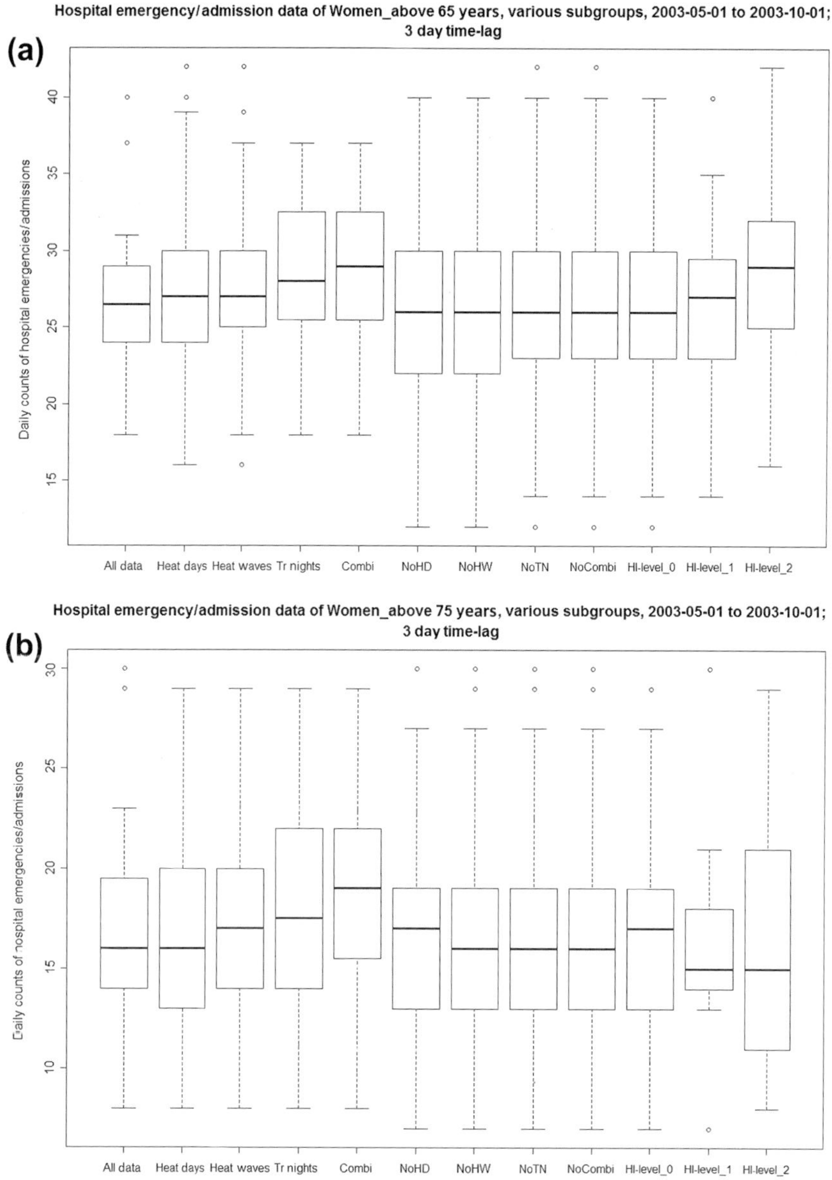

FIGURE 8.6 (a, b) The boxplots visualize the distribution of daily hospital emergency/admission counts for women (a) aged 65 years and older (the one above) and (b) aged 75 years and older (the one below) of the summer 2003 (May–September) according to various exposition factors with a time-lag of 3 days. The subgroup "Combi" (combination of days with heat wave and tropical night condition) has on average the most counts of emergencies/admissions in both plots. NoHD - Number of heat days (Number of heat days; i.e. days with maximum temperature ≥ 30 °C), NoHW - Number of heatwaves, i.e. number of events with three consecutive days of maximum temperature ≥ 30° C, NoTN - Number of tropical nights, i.e. days with minimum temperature not falling below 20 °C, NoCOMBI - Number of combined heat days and tropical nights.

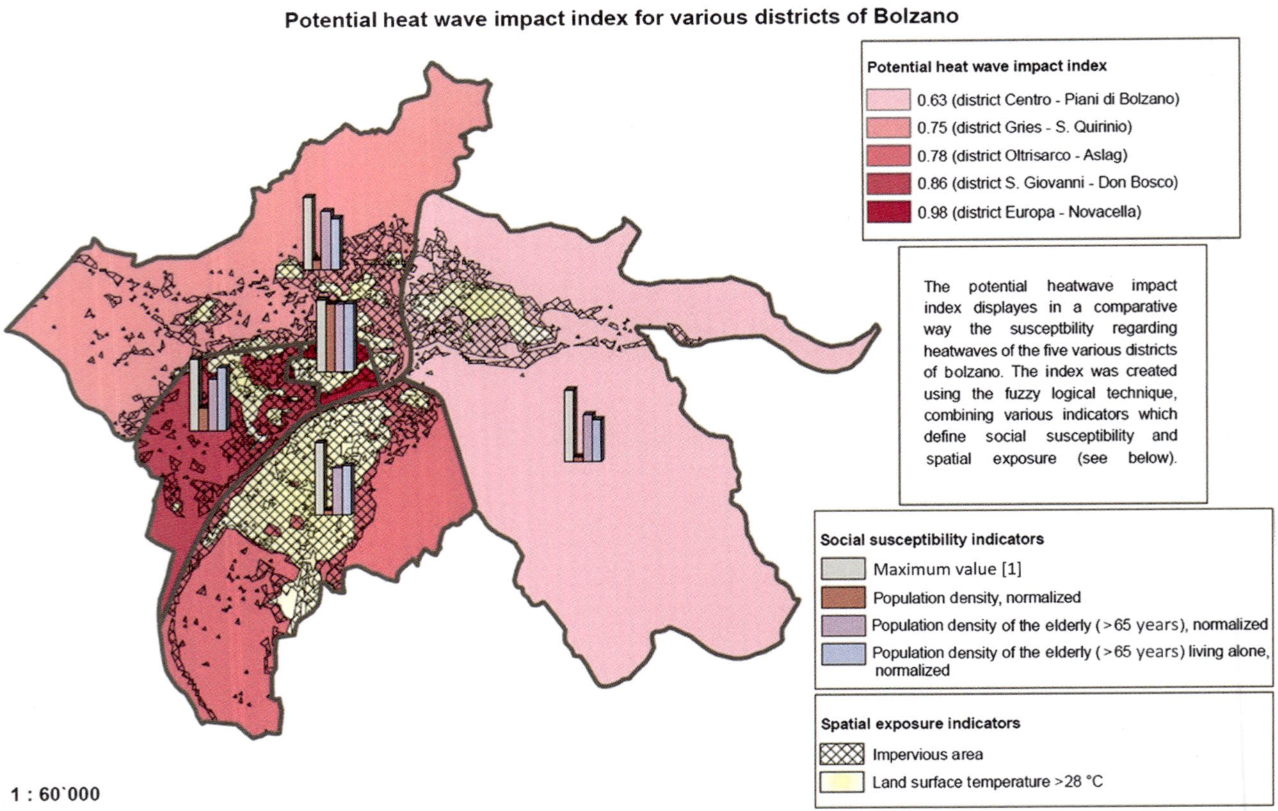

FIGURE 8.7 Potential heat wave impact index for the districts of Bolzano. The highest value was assigned to Europa/Novacella (0.98), the lowest to Centro/Piani di Bolzano (0.63). In between there are the districts S. Giovanni/Don Bosco (0.86), Oltrisarco/Aslag (0.78), and Gries/S. Quirinio (0.75).

8.5 CONCLUSIONS AND RECOMMENDATIONS

8.5.1 Landslides

A methodology for the development of a vulnerability curve for debris flow events based on information derived from past events is presented. The vulnerability curve can be used in order to assess future monetary loss in the area considering changes in the number of buildings involved. The inclusion of damages of future events in the development of the curve will increase its reliability and reduce uncertainties. Therefore, the quality of the damage information will play a major role in the reliability for the resulting curve. The results of the case study show clearly that there is a need for detailed documentation of the damages following a disastrous event. The vulnerability curve produced during the case study, together with a thorough damage documentation, can provide the stakeholders and decision makers with a tool for rapid damage assessment and loss estimation for future events. The documentation of damages and the vulnerability curve can provide information that can be used for vulnerability reduction of the elements at risk and reduction of loss due to landslides in the future.

8.5.2 Floods

Flood hazard zone maps show, in the current situation, a very high exposure for a large proportion of built-up areas of Sterzing and Pfitsch. The group discussion shows that participating residents are aware of flood risks, but assess the risk lower than flood management experts. Residents' capacity to anticipate is hampered by the long absence of floods. Residents assess risk especially for loss of life as low. Reliance on public protection measures and the warning system as well as positive experience of low damage in recent floods contribute to underestimating risks. Nevertheless, acceptance of the importance of new and better public flood protection measures is generally high among participants of the FGD. They support the reactivation of the "commission of civil protection" and putting flood defense higher on the political agenda.

Very low acceptance of self-precaution, low knowledge on private flood protection measures, and attitudes toward the split of responsibility—with participants often disclaiming their own responsibility—led to very few citizens performing private flood protection measures and therefore a low coping capacity at household scale.

8.5.3 Heat Waves

By means of a semiquantitative potential heat wave impact index, a spatial social vulnerability was shown for the five different districts of Bolzano. Hospital emergency/admission data and climate data were analyzed, but no stringent heat health issues could be shown for the elderly population of Bolzano. Nevertheless, some tendencies could be revealed which at least cannot negate a possible heat health issue due to the combination of heat days and tropical nights on elderly women. Considering future climate simulation, the frequency of heat waves will increase. In addition, the fraction of the very old people (>80 years)—hence the potentially most vulnerable—will increase due to increasing life expectancy. Therefore, it is most probable that the vulnerability related to heat wave issues in the city of Bolzano will increase and heat health impacts should be monitored more intensively in the near future.

At the moment, there are two differing heat wave warning systems at local and national level, which need to be better coordinated in order to avoid the confusion of the population. The detailed information required in order to analyze the situation in Bolzano needs extensive stakeholder involvement. Therefore, awareness of stakeholders and their interest in the investigations are key aspects for a successful study.

However, not only Bolzano but all areas of the province along the Adige River with altitudes below c. 600 m are affected by summer heat waves. Early warning systems and measures to protect those most vulnerable should be extended to those areas affected by heat waves beyond the city of Bolzano.

Furthermore, it is required to carry out long-term adaptation activities addressing urban planning issues such as the avoidance of urban heat islands and the extension of green areas.

A general recommendation is to increase awareness of the heat wave problematic, and the probable future aggravation of it, among policy makers, institutions, and citizens. This would build a stronger base and would raise acceptance for measures to mitigate and adapt to heat waves in Bolzano and the other affected areas of the Province.

8.5.4 Conclusions

South Tyrol is susceptible to a number of hazard types. Climate change is expected to influence in the future the magnitude and frequency of some events; however, the vulnerability of the population is also expected to change (more elderly people in the future) and should also be considered in future plans. Reducing vulnerability to natural hazards is the key to risk reduction. In this chapter, different dimensions of vulnerability were assessed by using a number of methods. The case study results in a number of recommendations, including better damage documentation for landslide events, development of long-term adaptation activities for more areas in south Tyrol regarding heat waves, and the enhancement of the coping capacity of citizens for flood events. The methods used in this chapter can be modified to be used for different types of hazards, but they can also be transferred to other places with similar hazard types.

REFERENCES

Autonome Provinz Bozen/Südtirol Hydrographisches Amt (APB/SHA), 2003. Climareport Extra Sommer 2003, E10.

Basu, R., Samet, J., 2002. An exposure assessment study of ambient heat exposure in the elderly population in Baltimore, Maryland. Environmental Health Perspectives 110 (12), 1213–1224.

Crozier, M.J., 1999. Slope instability: landslides. In: Alexander, D., Fairbridge, R.W. (Eds.), Encyclopaedia of Environmental Science. Dordrecht, pp. 561–562.

De Marchi, B., Scolobic, A., Delli Zotti, G., Del Zotto, M., 2007. Risk Construction and Social Vulnerability in an Italian Alpine Region. Task 11: Risk Perception, Community Behaviour and Social Resilience, Case Study Italy.

Drinkwater, B.L., Horvath, S.M., 1979. Heat tolerance and ageing. Medical Science and Sports 11 (1), 49–55.

Glade, T., Crozier, M., (2005) The nature of landslide hazard impact. In: Glade, T., Anderson, M,, Crozier, M,, (eds) Landslide hazard and risk. Wiley, Chichester, pp 43–74

IPCC, 2001. Climate change 2001: the scientific basis. In: Houghton, J.T., Ding, Y., Griggs, D.J., Noguer, M., van der Linden, P.J., Dai, X., Maskell, K., Johnson, C.A. (Eds.), Contribution of Working Group I to the Third Assessment Report of the Intergovernmental Panel on Climate Change. Cambridge University Press, Cambridge, United Kingdom and New York, NY, USA, 881 pp.

IPCC, (2012). Managing the risks of extreme events and disasters to advance climate change adaptation. A Special Report of Working Groups I and II of the Intergovernmental Panel on Climate Change [Field C.B., V., Barros, T.F., Stocker, D., Qin, D.J., Dokken, K.L., Ebi, M.D., Mastrandrea, K.J., Mach, G.-K., Plattner, S.K., Allen, M., Tignor, and P.M., Midgley, (eds.)]. Cambridge University Press, Cambridge, UK, and New York, NY, USA, , 582 pp

Kaswalder, C., 2009. Schätzungsstudie zur Berechnung des Schadenspotentials bei Hochwasserereignissen durch die Rienz im Abschnitt Bruneck-St. Lorenzen. Autonome Provinz Bozen, Südtirol.

Kropp, J., Holsten, A., Lissner, T., Roithmeier, O., Hattermann, F., Huang, S., Rock, J., Wechsung, F., Lüttger, A., Pompe, S.(U.F.Z.), Kühn, I.(U.F.Z.), Costa, L., Steinhäuser, M., Walther, C., Klaus, M., Ritchie, S., Metzger, M., 2009. Klimawandel in Nordrhein-Westfalen – Regionale Abschätzung der Anfälligkeit ausgewählter Sektoren. Abschlussbericht des Potsdam-Instituts für Klimafolgenforschung (PIK) für das Ministerium für Umwelt, Naturschutz, Landwirtschaft und Verbraucherschutz Nordrhein-Westfalen (MUNLV).

Papathoma-Köhle, M., Keiler, M., Totschnig, R., Glade, T., 2012. Improvement of vulnerability curves using data from extreme events: debris flow event in South Tyrol. Natural Hazards 64, 2083–2105.

Pfitscher, A., 1996. Wasserkatastrophen im Martelltal. Athesia, Bozen.

Röhrs, M., Glade, T., 2010. Südtirol. In: Bell, R., Mayer, J., Pohl, J., Greiving, S., Glade, T. (Eds.), Integrative Frühwarnsysteme für Gravitative Massenbewegungen (ILEWS), pp. 39–45

Schär, C., Vidale, P., Lüthi, D., 2004. The role of increasing temperature variability in Europe summer heat waves. Nature 427, 332–336.

Staffler, H., Pollinger, R., Zischg, A., Mani, P., 2008. Spatial variability and potential impacts of climate change on flood and debris flow hazard zone mapping and implications for risk management. Natural Hazards and Earth System Science 8, 539–558.

Steadman, R.G., 1979. The assessment of sultriness I. A temperature-humidity index based on human physiology and clothing science. Journal of Applied Meteorology 18, 861–873.

UNDRO (1984) Disaster prevention and mitigation—a compendium of current knowledge, vol 11. Preparedness Aspects, New York

Chapter 9

Conclusion: Assessing Vulnerability in Europe and the World

David E. Alexander

Institute for Risk and Disaster Reduction, University College London, London, United Kingdom

Chapter Outline

9.1 REDEFINING CONCEPTS AND MAKING CONNECTIONS

The word vulnerability comes from the Latin *vulnerare*, to wound or injure. This gave rise to the comparable term *vulnerate*, which entered the English language in the early seventeenth century, although it appears not to have mutated into *vulnerability* until 200 years later (*OED*). In more modern times, much attention has been given to the question of how to define vulnerability in relation to disasters (Weichselgartner, 2001; O'Brien and O'Keefe, 2013). In parallel, a substantial literature on vulnerability to poverty, often conceptualized in terms of failure of entitlement, has begged the question of the extent to which poverty and vulnerability to disasters are synonymous (Adger, 2006). The answer is that they are very nearly synonymous, although one must avoid the trap of circularity and also the tendency to think of all poor communities as vulnerable to disaster simply because they are poor. Moreover, vulnerability obviously exists in relation to phenomena other than disasters: unemployment and loss of livelihood and income, for example (Dutta et al., 2011). However, one result of the connection is that poverty is often treated as diagnostic of vulnerability to

Assessment of Vulnerability to Natural Hazards. http://dx.doi.org/10.1016/B978-0-12-410528-7.00009-6

disaster, as, along with health status, it is one of the best indicators of the general vulnerability of a social unit, be it a person, a family, a peer group, or a wider community.

Vulnerability is subject to a great paradox: it is a hypothetical concept, but one that nevertheless does not lack reality. It is simply not tangible: in the same way that in the physical world friction only comes into being when it is mobilized, so vulnerability is only physically apparent when it manifests itself as impact. This is one reason why the concept is difficult to measure. Confirmation of the existence of vulnerability is obtained post hoc by measuring the impact of disaster, or at least by inferring that past impacts will be diagnostic of future events. Vulnerability is thus a latent or inherent property of many different systems—physical, environmental, social, economic, cultural, institutional, and so on.

The fact that vulnerability is manifest in many different ways means that it can only adequately be understood as a multidimensional concept. In order to reduce the complexity to something understandable, it is tempting to restrict one's understanding to particular elements, for example, the physical aspects, and ignore the others. However, this is unwise because many of the most important aspects of vulnerability are those that pertain to the interactions. Hence, social vulnerability may be intimately linked to economic or institutional vulnerability, with feedbacks and mutual causality that are complex and challenging to study (Birkmann et al., 2013).

In seeking to understand vulnerability in the field, there is a risk of mistaken diagnosis. For example, in recent earthquakes in the Mediterranean region, there has been widespread collapse of both traditional masonry buildings and modern reinforced concrete ones (Liel and Lynch, 2012; Çelebi et al., 2013). While it is logical to attribute this to the vulnerability of buildings in terms of construction technique, properties of materials, state of maintenance, foundation performance, and so on, this is only part of the picture. In the L'Aquila (central Italy) earthquake of 6 April 2009, as in several recent seismic disasters in Turkey, the widespread and highly lethal collapse of modern reinforced concrete buildings, which should have been constructed with adequate earthquake resistance, also reflects the vulnerability of institutions, cultural norms, and social relations that allowed such a state of high susceptibility for damage to develop (Alexander, 2013a).

While vulnerability overlaps with poverty, it also does so with the concepts fragility, susceptibility, and exposure. Can a person, social entity, or physical asset be fragile but not vulnerable, or vice versa? The MOVE framework sees susceptibility and fragility as virtually synonymous and as being conditions under which vulnerability is manifest (MOVE Project, 2011). They are seen as descriptors of vulnerability states or dimensions. However, I would question the relationship between these terms. *Fragility* is a relative concept which implies that an entity can pass a threshold at which collapse, disintegration, deterioration, or failure become inevitable, or at least highly likely. As fragility is an inherent property of the entity, it is innate. *Susceptibility* is an external property which implies response to a stimulus. For instance, a building may not be susceptible to damage in small floods or earthquakes, but may be highly susceptible to collapse in large ones.

Exposure is a particularly difficult concept to pin down. In catastrophe insurance, it represents the full extent of an insurer's liability when there is a need to compensate a policy holder for the total loss of any insured asset (Crompton et al., 2013). However, exposure has other connotations in time and space. In this sense, an entity can be exposed but not vulnerable, if it suffers the effect of a hazard without significant losses being incurred. Activities expose people, organizations, and communities to risk in a transient, perhaps repetitive, manner. One of the best illustrations of this was first mapped out in relation to earthquakes in Chile, where it was recognized that the toll of human casualties will vary according to the timing of the earthquake in relation to aggregate patterns of human activity which put large numbers of people into different risk states, such as at home asleep, traveling, working in certain buildings, and so on (Lomnitz, 1970). The elusiveness of the concept of exposure means that it is often left out of the conceptual equation that links hazard with vulnerability to produce risk and disaster. In other words, it is subsumed into vulnerability, but never into hazard. A difficulty arises due to the fact that exposure can be viewed as an absolute (i.e., binary) or relative (i.e., variable) property; either an asset is exposed or not, or alternatively there may be degrees of exposure.

The definition of exposure is intimately linked to one of the central problems of disaster risk reduction: the relationship between magnitude and frequency (Cox, 2010). The general pattern is that high magnitude events have low frequency, and vice versa. For natural hazards, the time series of events is usually given as a straight line on lognormal scales. This implies a normal distribution, in which large events are equally as rare as small, frequent ones are insignificant. In reality, Hewitt (1970) long ago demonstrated that natural events obey ten or eleven different distributions. Nevertheless, the normal distribution has continued to be assumed in calculations of what will happen next (Keller et al., 1992). It is also assumed that there is little justification for preparing for major events that have a very low probability of occurrence, whatever their magnitude. Hence, society does not devote major resources to coping with an asteroid impact, even though sooner or later in Earth history one will occur.

There are two problems with the use of the normal distribution of magnitude–frequency relationships in managing hazards. The first is that large, low-frequency events may well be worth preparing for, whatever the statistics tell us. This is because the consequences are amenable to being reduced and are likely to be so serious that it makes sense to put mitigation measures in place, despite the rarity of the event. A further reason is that trending time series may make mitigation a progressively more attractive option. Hence, viewed over an adequate timescale, the distribution of all events may be both skewed and trending. In this case, we borrow from economics the concept of the "fat-tailed" distribution, in which high-magnitude events may be, or may become, more common than expected (Nordhaus, 2011). This is especially the case for meteorological disasters, which will undoubtedly respond to changes in climate that make extreme events more frequent, and possibly more violent. Given the nonlinear relationship between wind speed and damage (Leatherman, 2011), this means that preparedness will need to be strengthened disproportionately if impacts are to be contained.

If, in all its connotations, vulnerability represents the negative side of susceptibility to disasters, what represents the positive; or, in other words, what is the opposite of vulnerability? Generally, the answer is an amalgam of the terms capability, capacity, coping, and resilience. The first three represent the surplus resources of society, in both a static and a dynamic sense. The last term is more problematic. In more than two millennia of history, resilience has enjoyed many different definitions and, since the concept began to be used in science during the seventeenth century, it has spread in a capillary manner from one discipline to another (Alexander, 2013b). Perhaps its best definition in relation to disasters comes from the analogy with mechanics, a discipline in which it has been used with considerable success. A resilient material must have an optimum combination of strength and ductility. Hence, the strength of society lies in its ability to resist disaster. This concept has been operationalized in the concept of *hardening* (Hyslop and Collins, 2013). Society's "ductility" exists in its *adaptability* (Handmer, 2003). This is a fundamental concept in terms of a century of development, by geographers and anthropologists, of the human ecology of adaptation to environmental hazards (Burton et al., 1968).

Inherent in the concepts of capability and resilience is the idea of social capital. It has been used as a means of identifying the capacity to mitigate and recover from disasters (Murphy, 2007; Aldrich, 2012). However, it has also been subject to sharp criticism. Thus, Haynes (2009) argued that social capital is neither social nor capital. He suggested that the concept lacks coherence, and that it cannot adequately be measured, especially in terms of any causality inherent in it. Moreover, it has a dark side, in that not all social capital works to the general benefit of society. As a result, he argued, it is difficult to operationalize the concept in any meaningful way. This has not deterred authors from trying (e.g., Brunie, 2010).

Disaster studies have long been dogged by disagreements over the definition of terms. Hence, O'Brien and O'Keefe (2013, pp. 129–131) listed 28 definitions of "resilience", and lists such as these have a tendency to grow as more and more opinions, experiences, and perspectives accrete. As a result of the controversies, it seems unwise to list a single antonym for vulnerability to disasters. Resilience can be created without reducing vulnerability. All it requires is attention to the wrong issues. It may, for example, finish the day with intact roads but collapsed bridges, or a population that survives, but key institutions that do not. Similarly, a community may be able to mitigate or cope with disaster, but still manifest a high level of vulnerability.

It is often assumed that vulnerability must be reduced at the level of communities. Community involvement in disaster risk reduction is seen by many as a panacea, or at least a primary goal (Berkes and Ross, 2013). Although there is some truth in this, there are several fundamental problems. One is the difficulty of defining what exactly a community consists, and at what scale it operates. To have any meaning, the concept of community must involve a significant level of positive interaction between participants. In an age distinguished by the primacy of information and communications technology, this does not have to involve spatial proximity. Hence, communities can be distributed among locations and may hang together by virtue of their common interests. In this respect, so-called "virtual" communities are collections of individuals that have common interests, not groups of people who find themselves living in close proximity with one another, or are constrained to do so.

In terms of defending territories against disaster impacts, a community will obviously be localized and centralized. This begs the question of what is its geographical scale, a region, city, neighborhood, street, or what? The matter is complicated by the fact that communities, if they exist, are not homogeneous entities. This is true, not merely in terms of their components, which are divided by race, class, culture, gender, and so on, but also with respect to possibly conflicting goals and objectives within a community. In development studies, much attention has been given to the phenomenon of *elite capture* (e.g., Kundu, 2011), which is an expression of the fact that a community's goals are often an expression of the will of its most powerful members. This seems to be true in most kinds of society, rich and poor, and it militates against designing effective vulnerability reduction measures by employing participatory methods and transparent governance (Wisner, 2013).

When disaster strikes, the theater of operations for managing the response to its impact is always local. If one has faith in the idea that a primary vehicle for tackling the problem is the community, then this must be conceptualized at a local scale. "Local" here can mean the scale of thousands, hundreds, or tens of inhabitants, but the effectiveness of community among them depends on the intensity and quality of interaction between key members (Boon et al., 2012). Thus, in a community at least some members must presumably know each other personally and most or all must have some sense of belonging. There must be some feeling of *genius loci*, or sense of place, with which to identify. Generally, dysfunctional communities will have high vulnerability to disaster at the community level. They may be defined by the homogeneity of their susceptibility, if by no other form. It is difficult to put any faith in the concept of community if most members are there under sufferance, for example, as a result of forced migration or utter lack of alternatives.

9.2 BRIDGING THE HUMAN–PHYSICAL GAP

Studies of disaster essentially began with attempts to understand hazard and its impact. The dominant position of the physical sciences over the social disciplines ensured that this was inevitable (Hewitt, 1983). Throughout the last 100 years, physical science studies of hazard and disaster have always been endowed with better resources and have usually had a sharper focus than the social disciplines have managed to acquire. One ends up with the absolute primacy of studies of hazard over those of vulnerability. Moreover, there is a greater propensity to study physical vulnerability than social aspects of the problem. Even in psychology, which is the one social discipline to have produced a copious literature on disasters, the vast majority of studies refer to postdisaster states and interventions, not predisaster susceptibility.

By placing society within environment, the MOVE framework aims to rectify this situation, and also to bridge the gap between the two views of disaster, one in which the primary cause is the physical hazard, and the other in which the hazard is merely the trigger for agents that act upon vulnerability (Hewitt, 1983).

One connection between the physical and social worlds is inherent in socio-natural events, including the so-called "na-techs", natural disasters with a technological component or natural events that disrupt technology such that it is the latter that causes the disaster (Santella et al., 2011). These fall into the category that involves composite, compound, and cascading disasters. For example, the earthquake of 11 March 2011 in the ocean east of Japan caused a millennial tsunami and led to the release of nuclear radiation in the Fukushima reactor meltdowns (Shadrina, 2012). Casualties and damage caused by the earthquake were limited: those resulting from the tsunami were exceptionally large (18,500 people killed and 500 km^2 of coastal land devastated by the waves). However, it may be that the Fukushima nuclear emergency was the biggest of the three concurrent disasters, as the wide dispersion of radioactivity, the persistence of radiation contamination, and the high costs of cleanup may eventually outweigh the other effects: time will tell. At the very least, in Europe, as elsewhere in the world, Fukushima has led to a very public increase in risk aversion regarding nuclear power generation (Gallucci, 2012). Pure risk aversion seeks to reduce vulnerability by eliminating it. The slightly less pure version creates high reliability systems by endowing them with robust mechanisms and considerable redundancy, at high cost, of course.

These reflections indicate that physical and human factors can be bound together intimately in terms of the primary impact of disaster. Failure to consider disasters as holistic problems is one reason why the two sides have traditionally been separate in the analysis of consequences, and this is a reflection of the narrowness of the specialties of researchers. It also reflects the fact that many technical officers, whose task is to remedy the effects or reduce the risk of disaster, only operate in a defined field, for example, structural engineering, and have little incentive to take into account other aspects of the problem, such as the way buildings are perceived and used by their occupants. Getting such people to recognize the importance of a wider view is a difficult but essential task. Not only mindsets and perceptions are involved but also institutional constraints and incentives, or disincentives.

9.3 THE COMPONENTS OF VULNERABILITY

The ability to disaggregate vulnerability into different components indicates that it can take different forms. The field of business continuity management (BCM) is founded on the notion that vulnerability has individual components that apply, for example, to the supply chain, the manufacturing or production process, relationships with clients, customers and suppliers, the market value of a company's shares, its position with respect to competitors, and its reputation with investors and customers. Damage in any of these sectors can activate vulnerability in other categories (Kemp, 2007). Hence, in BCM, vulnerability analysis tends to be conducted by sector. In other fields of analysis it is considered by discipline: ecology, economics, social studies, engineering, anthropology, and so on. Once again, this tends to minimize the role of the all-important connections between these realms. For example, it is easy to think of examples in which the disaster resulted largely from failure to enact and enforce construction norms, so that lethal damage was eventually caused by extreme natural processes: lack of knowledge might not be the problem, but it is instead failure to utilize it, hence the question is one of the politics of engineering, possibly also the ethics (Walker, 2010).

If vulnerability must be reduced to components, this should be done with respect to its dynamism. One possible interpretation is that it can be defined relative to the circumstances that generate it. The following model breaks vulnerability down according to its context (Alexander, 1997; Özerdem and Jacoby, 2006):

- *Total vulnerability:* life is generally precarious because little or nothing has been done to reduce the sources and potential impacts of risk. This condition tends to apply to poor and marginalized societies which lack the resources to protect themselves.
- *Economic:* people lack adequate occupation and hence vulnerability refers to the precariousness of their productive activities and sources of income.
- *Technological or technocratic:* caused by the riskiness of technology or of the ways in which it is utilized.
- *Residual:* caused by lack of modernization, in which risk conditions evolve but mitigation strategies do not keep pace with them.
- *Delinquent:* caused by corruption, negligence, or criminal activity that puts people or assets at risk.
- *Newly generated:* caused by change in circumstances, for example, as a result of newly emerging risks.

Such is the complexity of society that these categories need not be mutually exclusive.

If vulnerability takes different forms or has different components not only will it be multifaceted but there will also be interaction between the components. Hence, it may be additive and there may be *Gestalt* (i.e., the whole may be more than the sum of the parts). Chains of causality, interactions between the parts, and collateral and secondary vulnerabilities may come into play. Thus we may define primary vulnerability as the direct product of cause and effect. For example, if an earthquake shakes a dilapidated house the poor quality of the masonry may cause the building to collapse. No gestalt is present. Secondary vulnerability, potentially with moderate *Gestalt*, results from the interaction of causes or the occurrence of coincidences. For instance, a building may resist earthquake shaking but not the water wave caused by the seismically induced breaching of a dam upstream. Complex vulnerability involves high *Gestalt* and occurs when complicated interactions between components heighten overall vulnerability. The ramified economic effects of a major earthquake in a metropolitan city would produce this (Menoni, 2001).

Vulnerability can be measured or estimated directly as potential for harm or loss. However, this requires hypotheses of the potential impact of events that have not occurred but are likely to. There is, of course, scope for inaccuracy in such theorizing. It can also be measured indirectly as "nonresilience"—i.e., the inability to be robust in the face of a threatening event. The fault trees and event trees used in the risk analysis of industrial processes are an example of this (Fischhoff et al., 1978; You and Tonon, 2012) and could be adapted to other forms of investigation, for example with society.

Like risk, vulnerability can be chronic or catastrophic, depending on whether it refers to the potential for a widely diffused malaise or a concentrated disaster. It applies to known, stable risks, mutating risks, emerging risks, and unknown risks. Whereas the process of investigating vulnerability in the context of the first of these is relatively straightforward, it becomes much more difficult when applied to adapting risks (such as those associated with the occurrence of more extreme events due to climate change) and emerging risks (e.g., new strains of pandemic disease). It cannot be achieved for unknown risks, the ones that will emerge in future scenarios. The management of vulnerability must involve holistic techniques that take account of the categories in which it occurs and the different degrees and levels of interaction between them (Cardona, 2004). This is necessarily so, as managing vulnerability with regard to single themes and cases will merely allow it to escape into other categories and will thus not reduce it. Holistic analyses must precede holistic remedies (McEntire, 2001).

Vulnerability is susceptible to the dialectic of forces that act to increase or diminish it. The concept is not static but is continually modified by forces that amplify or mitigate it. Thus the vulnerability of settlement on a floodplain can be diminished by structural and planning measures but simultaneously amplified by building new structures that are at risk of inundation (Scheuer et al., 2011). Perception is the deciding factor, providing one takes into account its links with decision-making and action. Low or inaccurate perception of hazard can perpetuate vulnerability, while high perception can lead to its reduction. However, it does not follow that strong and accurate perception will necessarily be good for disaster risk reduction. That depends on what other factors exist in society to reduce or perpetuate vulnerability (Heijmans, 2001).

Given the dynamic environment in which it occurs, measures taken to reduce vulnerability need to be sustainable. If they are not, it will inevitably climb back again under the pressure of socioeconomic opportunism. Sustainable vulnerability reduction is locally based, supported by the community, well integrated into legislation and planning instruments, and part of a grand strategy to make life more resilient for the inhabitants of the area in question. It goes without saying that resilience, too, depends on sustainability (Wisner, 2004).

9.4 VALEDICTION OF THE MOVE FRAMEWORK

Many of the problems and opportunities discussed above were evaluated and tackled in the MOVE project. The first objective was to create a framework that linked the disparate elements of vulnerability so that they could be studied in a unified way, and one in which the elements of vulnerability could be selected for study in relation to the demands of the problem under investigation. The second objective was to show that results of use to practitioners could be derived from such an exercise. The main categories of practitioner included urban and regional planners, emergency planners, environmental decision makers, people responsible for monitoring and reducing natural hazards, forecasters and civil protection managers. Most of these were in the public sector, but there was also a need to involve private-sector managers, for example, from among the operators of critical infrastructure.

In almost a century of academic activity, disaster studies have evolved into a sort of "interdisciplinary discipline", one that involves at least 35 professions and well-defined areas of study. The field is, if not unique, at least highly distinctive in that the body of theory it produces needs to be immediately applicable to practical problems. The value of theory is that it illuminates complex situations and makes them intelligible by relating them to a body of knowledge derived from kindred circumstances elsewhere, or in other words by measuring them against known regularities. The chaos produced by disaster impacts, and the complexity of multidimensional vulnerabilities are thus made intelligible, or at least susceptible to valid interpretation, by the application of theory.

The MOVE framework was tested in a variety of European environments, from mountains to plains and coasts, from rural to metropolitan areas. Moreover, it was used in the context of a variety of methodologies, from remote sensing to geographic information systems, econometrics, time-series analysis, scenario-building, and so on. The framework is, of course, a model, and the essence of "model" is that it be an elegant simplification of reality, one that removes extraneous detail in order to uncover the fundamental details of a relationship and, by simplifying them, make them intelligible. Any framework that has to simplify in order to be used in many physical and social environments, and with many different techniques, risks having one of the principal defects of such a model: that it is too general to have much explanatory power.

However, a framework is a special kind of model. It does not need to explain phenomena and links very deeply, but it should provide the guidelines by which the techniques applied to data and situations can bring to light causal mechanisms, *in such a way that they can profitably be related to each other*. The *Gestalt* in the MOVE framework is that, by providing a network of facets or components (social, physical, resilience, fragility, risk reduction, etc.) it facilitates a common approach, in which the elements are used selectively. Causality emerges from both the linkages between elements and subsystems in the framework (e.g., social and institutional, fragility and susceptibility, risk governance, and adaptation), and by comparing different situations using the same basic methodology. The framework also demonstrates that vulnerability has both an essential unity that transcends geographical factors, but also different shades of meaning at various spatial scales. Hence, national vulnerability (the threat to a national way of life, or, in extremis, to the integrity of a nation) is not quite the same thing as local vulnerability, which compromises local life in a way that is, perhaps, remediable at broader spatial scales.

This book has made it abundantly clear that vulnerability is a complex, dynamic phenomenon which exhibits many contrasting facets and interacts with processes such as risk reduction, adaptation, and coping. The MOVE framework, and the studies in which it has been used, show that the concept of vulnerability also has an essential unity which transcends spatial and temporal scales, and is inherent in many different phenomena, both physical and social. In writing this book, we hope that vulnerability studies will be given more weight in assessments of hazard impacts. Despite significant progress (Birkmann, 2013), this remains a major challenge as we struggle to understand the complexity of disasters so that we can subdue them and help create a safer world for humanity to live in.

REFERENCES

Adger, N., 2006. Vulnerability. Global Environmental Change 16, 268–281.

Aldrich, D.P., 2012. Building Resilience: Social Capital in Post-Disaster Recovery. University of Chicago Press, Chicago, 256 pp.

Alexander, D.E., 1997. The study of natural disasters, 1977-97: some reflections on a changing field of knowledge. Disasters 21 (4), 284–305.

Alexander, D.E., 2013a. An evaluation of the medium-term recovery process after the 6 April 2009 earthquake in L'Aquila, Central Italy. Environmental Hazards: Human and Policy Dimensions 12 (1), 60–73.

Alexander, D.E., 2013b. Resilience and disaster risk reduction: an etymological journey. Natural Hazards and Earth System Sciences 13 (11), 2707–2716.

Berkes, F., Ross, H., 2013. Community resilience: toward an integrated approach. Society and Natural Resources 26 (1), 5–20.

Birkmann, J. (Ed.), 2013. Measuring Vulnerability to Natural Hazards: Towards Disaster Resilient Societies, second ed. United Nations University Press, Tokyo, 686 pp.

Birkmann, J., Cardona, O.D., Carreño, M.L., Barbat, A.H., Pelling, M., Schneiderbauer, S., Kienberger, S., Keiler, M., Alexander, D., Zeil, P., Welle, T., 2013. Framing vulnerability, risk and societal responses: the MOVE framework. Natural Hazards 67 (2), 193–211.

Boon, H.J., Cottrell, A., King, D., Stevenson, R.B., Millar, J., 2012. Bronfenbrenner's bioecological theory for modelling community resilience to natural disasters. Natural Hazards 60 (2), 381–408.

Brunie, A., 2010. Household awareness of what to do in a disaster: a social capital approach. International Journal of Mass Emergencies and Disasters 28 (1), 59–86.

Burton, I., Kates, R.W., White, G.F., 1968. The Human Ecology of Extreme Geophysical Events. Working Paper no. 1. Natural Hazards Research and Applications Information Centre. University of Colorado, Boulder, Colorado, 37 pp.

Cardona, O.D., 2004. The need for rethinking the concepts of vulnerability and risk from a holistic perspective: a necessary review and criticism for effective risk management. In: Bankoff, G., Frerks, G., Hilhorst, D. (Eds.), Mapping Vulnerability: Disasters, Development and People. Earthscan, London, pp. 37–51.

Çelebi, E., Aktas, M., Çağlar, N., Özocak, A., Kutanis, M., 2013. October 23, 2011 Turkey/Van-Ercis earthquake: structural damages in the residential buildings. Natural Hazards 65 (3), 2287–2310.

Cox Jr., L.A., 2010. The impossibility of defining "frequency" satisfactorily, revisited. Risk Analysis 30 (10), 1464–1465.

Crompton, R.P., McAneney, K.J., Musulin, R.T., Pielke Jr., R.A., 2013. Natural disaster losses and climate change. Safety, reliability, risk and life-Cycle performance of structures and infrastructures. In: Proceedings of the 11th International Conference on Structural Safety and Reliability, ICOSSAR 2013, pp. 1455–1461.

Dutta, I., Foster, J., Mishra, A., 2011. On measuring vulnerability to poverty. Social Choice and Welfare 37 (4), 743–761.

Fischhoff, B., Slovic, P., Lictenstein, S., 1978. Fault trees: sensitivity of assessed failure probabilities to problem representation. Journal of Experimental Psychology: Human Perceptions and Performance 4, 330–344.

Gallucci, R., 2012. What—me worry? Why so serious? A personal view on the Fukushima nuclear reactor accidents. Risk Analysis 32 (9), 1444–1450.

Handmer, J., 2003. Adaptive capacity: what does it mean in the context of natural hazards? In: Smith, J.B., Klein, R.J.T., Huq, S. (Eds.), Climate Change, Adaptive Capacity and Development. Imperial College Press, London, pp. 51–70.

Haynes, P., 2009. Before Going Any Further with Social Capital: Eight Key Criticisms to Address. INGENIO (CSIC-UPV) Working Paper Series 2009/02, Institute of Innovation and Knowledge Management. Polytechnic University of Valencia, Valencia, 22 pp.

Heijmans, A., 2001. Vulnerability: A Matter of Perception. Working Paper no. 4, Benfield Grieg Hazard Research Centre. University College London, London, 17 pp.

Hewitt, K., 1970. Probablistic approaches to discrete natural events: a review and theoretical discussion. Economic Geography Supplement 46 (2), 332–349.

Hewitt, K., 1983. The idea of calamity in a technocratic age. In: Hewitt, K. (Ed.), Interpretations of Calamity from the Viewpoint of Human Ecology. Unwin-Hyman, London, pp. 3–32.

Hyslop, M.P., Collins, A.E., 2013. Hardened institutions and disaster risk reduction. Environmental Hazards 12 (1), 19–31.

Keller, A.Z., Wilson, H.C., Al-Madhari, A., 1992. Proposed disaster scale and associated model for calculating return periods for disasters of given magnitude. Disaster Prevention and Management 1 (1), 26–33.

Kemp, R.L., 2007. Vulnerability assessments for public and private facilities. Journal of Business Continuity and Emergency Planning 1 (3), 245–251.

Kundu, D., 2011. Elite capture in participatory urban governance. Economic and Political Weekly 66 (10), 23–25.

Leatherman, S.P., 2011. Hurricane wind damage mitigation: research and outlook. Natural Hazards Review 12 (4), 202–206.

Liel, A.B., Lynch, K.P., 2012. Vulnerability of reinforced-concrete-frame buildings and their occupants in the 2009 L'Aquila, Italy, earthquake. Natural Hazards Review 13 (1), 11–23.

Lomnitz, C., 1970. Casualties and behaviour of populations during earthquakes. Bulletin of the Seismological Society of America 60 (4), 1309–1313.

McEntire, D.A., 2001. Triggering agents, vulnerabilities and disaster reduction: towards a holistic paradigm. Disaster Prevention and Management 10 (3), 189–196.

Menoni, S., 2001. Chains of damages and failures in a metropolitan environment: some observations on the Kobe earthquake in 1995. Journal of Hazardous Materials 86 (1–3), 101–119.

MOVE Project, 2011. Assessing Vulnerability to Natural Hazards in Europe: From Principles to Practice. A Manual on Concept, Methodology and Tools. MOVE: Improvement of Methods for Vulnerability Assessment in Europe, Florence, Italy, 48 pp.

Murphy, B.L., 2007. Locating social capital in resilient community-level emergency management. Natural Hazards 41 (2), 297–315.

Nordhaus, W.D., 2011. The economics of tail events with an application to climate change. Review of Environmental Economics and Policy 5 (2), 240–257.

O'Brien, G., O'Keefe, P., 2013. Managing Adaptation to Climate Risk: Beyond Fragmented Responses. Routledge, London, 217 pp.

Özerdem, A., Jacoby, T., 2006. Disaster Management and Civil Society: Earthquake Relief in Japan, Turkey and India. International Library of Postwar Reconstruction and Development no. 1. I.B. Taurus, London and New York, 142 pp.

Santella, N., Steinberg, L.J., Aguirra, G.A., 2011. Empirical estimation of the conditional probability of natech events within the United States. Risk Analysis 31 (6), 951–968.

Scheuer, S., Haase, D., Meyer, V., 2011. Exploring multicriteria flood vulnerability by integrating economic, social and ecological dimensions of flood risk and coping capacity: from a starting point view towards an end point view of vulnerability. Natural Hazards 58 (2), 731–751.

Shadrina, E., 2012. Fukushima fallout: gauging the change in Japanese nuclear energy policy. International Journal of Disaster Risk Science 3 (2), 69–83.

Walker, P., 2010. Reflection: from science to politics and back again. Environmental Hazards 9 (4), 395–398.

Weichselgartner, J., 2001. Disaster mitigation: the concept of vulnerability revisited. *Disaster Prevention and Management* 10 (2), 85–94.

Wisner, B., 2004. Assessment of capability and vulnerability. In: Bankoff, G., Frerks, G., Hilhorst, D. (Eds.), Mapping Vulnerability: Disasters, Development and People. Earthscan, London, pp. 183–193.

Wisner, B., 2013. Self-assessment of coping capacity: participatory, proactive, and qualitative engagement of communities in their own risk management. In: Birkmann, J. (Ed.), Measuring Vulnerability to Natural Hazards: Towards Disaster Resilient Societies, second ed. United Nations University Press, Tokyo, pp. 440–454.

You, X., Tonon, F., 2012. Event-tree analysis with imprecise probabilities. Risk Analysis 32 (2), 330–344.

Index

Note: Page numbers followed by *"f"* indicate figures; *"t"* tables.